RAND NATIONAL DEFENSE RESEARCH INSTITUTE

Reforming Military Retirement

Analysis in Support of the Military Compensation and Retirement Modernization Commission

Beth J. Asch, Michael G. Mattock, James Hosek

Prepared for the Office of the Secretary of Defense

Approved for public release; distribution unlimited

For more information on this publication, visit www.rand.org/t/rr1022

Library of Congress Control Number: 2015944863
ISBN: 978-0-8330-9015-7

Published by the RAND Corporation, Santa Monica, Calif.

Cover image by Staff Sgt. Justin Shemansk.

Preface

The National Defense Authorization Act of 2013 mandated an independent commission—the Military Compensation and Retirement Modernization Commission (MCRMC)—to review the military compensation and retirement systems and make recommendations to modernize them. The Commission's pay and retirement working group asked RAND to provide analytic support to inform their deliberations. In recent years, RAND has developed a modeling capability, known as the Dynamic Retention Model (DRM), that models the retention decisions of military members. The model can be used to simulate the effects of alternative military compensation reforms on retention, personnel costs, and government outlays. This capability was most recently used to support the deliberations of an internal Department of Defense (DoD) working group on compensation reform, documented in the RAND report *Toward Meaningful Military Compensation Reform: Research in Support of DoD's Review of Military Compensation* (Asch, Hosek, and Mattock, 2014). MCRMC asked RAND to adapt the DRM to simulate personnel retention, cost, and government outlays under alternative MCRMC proposals.

This report documents the research conducted for this effort. It summarizes the steady-state retention effects in the active and reserve components for each service—for enlisted personnel and officers—of the MCRMC retirement reform proposal. In addition, it shows costs in the steady state as well as costs and outlays for active component personnel in the transition years to the steady state. This document should be of interest to those concerned with military compensation reform and those specifically interested in the analytic underpinnings of the retirement reform proposals offered by the MCRMC.

This research was conducted within the Forces and Resources Policy Center of the RAND National Defense Research Institute, a federally funded research and development center sponsored by the Office of the Secretary of Defense, the Joint Staff, the Unified Combatant Commands, the Navy, the Marine Corps, the defense agencies, and the defense Intelligence Community.

For more information on the Forces and Resources Policy Center, see http://www.rand.org/nsrd/ndri/centers/frp.html or contact the director (contact information is provided on the web page).

Contents

Figures

Tables

Summary

The Military Compensation and Retirement Modernization Commission (MCRMC) retirement reform plan is a blended approach that includes a defined benefit (DB), a defined contribution (DC) plan, and higher current compensation in the form of continuation pay at year of service (YOS) 12. In addition, the plan allows active component (AC) service members—at the time of their retirement from the military—a choice regarding the DB annuity receivable from the time of retirement to age 67.[1] The member may choose a full DB annuity, a full lump-sum payment in lieu of the annuity, or partial DB annuity and partial lump sum. The DB is like today's DB, except with a multiplier of 2.0 percent instead of 2.5 percent in today's system. Reserve component (RC) retirees could elect to receive (1) a full DB annuity starting at age 60; (2) a lump sum paid at the time of retirement from the RC in lieu of the annuity to age 67; or (3) a partial DB annuity from ages 60 to 67 and a partial lump sum paid at the time of retirement from the RC, then receive the full annuity starting at age 67. The DC plan vests at YOS 3, the Department of Defense (DoD) makes an automatic contribution of 1 percent of basic pay from years 1 to 20, and DoD matches the member's contribution up to 5 percent of basic pay over years 3–20.

The National Defense Authorization Act of 2013 required DoD to transmit its recommendations for compensation reform to the MCRMC and for MCMRC to conduct public hearings on DoD's recommendations. DoD transmitted two concepts for compensation reform, both of which were blended plans, and the MCRMC's blended plan can be seen as its response to the DoD concepts. Similar to the DoD concepts in certain ways but different in others, the MCRMC plan offers innovations that add value to the current compensation package by allowing retiring service members greater choice over the timing and amount of benefit payments, strengthening the incentive to save for the future via the DC plan, providing the services with more-flexible force management and career shaping by occupation, and generating cost savings.

MCRMC engaged the RAND National Defense Research Institute for analytical support during its internal deliberations regarding the form and details of its retirement plan. We based our analysis on the RAND Dynamic Retention Model (DRM), a dynamic programming model of individual choice regarding AC retention and RC participation that has been estimated based on longitudinal data and with significant capability to simulate alternative compensation policies. An important criterion of the RAND analysis was whether a reform

[1] The MCRMC proposal set the age of full retirement benefits at the age of full Social Security benefits. In our modeling, we set this age to 67. In the remainder of the document, we refer to age 67 as the age of full retirement, with the understanding that the proposal is actually the age of full Social Security benefits.

could sustain the current force size and shape. We found that the MCRMC plan could do so; this was the case by service, for officer and enlisted, for AC and RC.

Further, the MCRMC plan would decrease cost. We estimated cost savings in the steady state of $2.3 billion per year to $7.7 billion per year, depending on the DC match rate and the lump sum versus second-career annuity choice, with an intermediate example showing cost savings of $4.3 billion per year. Our estimate differs from that reported in the MCRMC final report (MCRMC, 2015) for several reasons, including that our estimate includes the cost savings to both DoD and the Department of the Treasury and includes the cost savings associated with the RC.

We also simulated the MCRMC plan for the transition years from implementation to steady state. The 2013 National Defense Authorization Act stipulated that currently serving members and retired members are grandfathered under the current system but allowed those members the choice to enter the new system (i.e., to opt in). During the transition years and opt in, the MCRMC plan sustained force size and shape from year to year and generated immediate cost savings in the form of lower retirement accrual charges. It also led to an increase in government outlays in the initial years to pay for DC contributions, continuation pay, and lump-sum payments for retirees choosing a lump-sum option, and then outlays decreased below baseline and aligned with cost savings in the long run.

Thus, we found that the MCRMC plan was effective in meeting manning requirements, cost-effective in being able to generate the same force size and shape at lower cost, valuable to the service member because of the early-vesting DC and the lump-sum choice, and valuable to the military services by offering the potential for greater flexibility in force management while still retaining a large portion of compensation in the DB plan. Past studies and commissions have concluded that blended plans hold the promise of greater efficiency, equity, and flexibility than the current compensation system. The MCRMC plan, as a blended plan, holds similar promise.

Acknowledgments

We gratefully acknowledge the helpful comments, guidance, and stimulating engagement of our colleagues at the MCRMC: Robert Daigle, Moira Flanders, Lyle Hogue, Steve Cylke, Edna Falk Curtin, and Derek Vestal. In addition, Joel Sitrin, Peter Zouras, and Rich Allen of the DoD Office of the Actuary cheerfully provided key input for our analysis, no matter the time pressure. We also thank our RAND colleagues David Knapp and Jennifer Lewis for their insights at crucial points in this project. Finally, we are grateful to Kathleen Mullen of RAND and Matthew Goldberg of the Congressional Budget Office, who provided helpful reviews and comments on an earlier draft.

Abbreviations

AC	active component
CRDP	Concurrent Retirement and Disability Pay
CRSC	Combat Related Special Compensation
DB	defined benefit
DC	defined contribution
DoD	Department of Defense
DRM	Dynamic Retention Model
FY	fiscal year
MCRMC	Military Compensation and Retirement Modernization Commission
QRMC	Quadrennial Review of Military Compensation
RC	reserve component
VA	Department of Veterans Affairs
WEX	Defense Manpower Data Center Work Experience File
YOS	year of service

CHAPTER ONE

Introduction

The National Defense Authorization Act for 2013 established an independent commission, known as the Military Compensation and Retirement Modernization Commission (MCRMC), "to review and make recommendations to modernize the military compensation and retirement systems." The Commission was chartered by law to conduct this review "in the context of all elements of the current military compensation and retirement systems, force management objectives, and changes in life expectancy and the labor force." The Commission was further directed by the President to focus on benefits related to compensation, including health care and family programs, along with various other topics according to principles of compensation and retirement modernization established by the President and transmitted to the Commission.

The establishment of a commission to consider military compensation reform follows a long history of commissions and study groups aimed at recommending changes to the compensation system, especially the retirement system. Indeed, the President generally directs the Department of Defense (DoD) to review its compensation system every four years, as part of the so-called Quadrennial Review of Military Compensation (QRMC). The role of the MCRMC is critical to informing the current policy discussion of military compensation, and the President has decided not to convene the usual QRMC this time, noting that the MCRMC report will serve an equivalent purpose. Furthermore, DoD issued a white paper in March 2014 on concepts for modernizing the military retirement system following an 18-month review of the system (U.S. Department of Defense, 2014), thereby providing DoD input to the MCRMC deliberations.

The impetus for the MCRMC comes from the ongoing importance of ensuring that the nation's service members are paid adequately and fairly; that their families are supported through programs aimed at housing, children's education, and spouse employment and education, among others; and that members and families have access to high-quality, affordable health care. Equally important is the role of compensation in ensuring that the nation has a fully manned and highly capable fighting force, and that the resources provided are sufficient to accomplish this without reaching a point of excess. Beyond defense per se is the national concern about controlling the growth of the federal budget, the deficit, and national debt.

Past commissions and studies, while appreciating the strengths of the military compensation system, have identified its deficiencies, particularly deficiencies of the military retirement system. The current military retirement system is a defined-benefit (DB) plan: The amount of the benefit is derived from a formula based on basic pay, years of service (YOS), and a multiplier. Service members are vested at 20 YOS and, in the case of active component (AC) members, receive an immediate annuity upon separation. Reserve component (RC) members with

20 qualifying years generally begin receiving retired pay at age 60. The retirement system has existed in its basic form for nearly 70 years.

The main criticisms are that the system is inflexible, inefficient, and inequitable. It is inflexible because of its one-size-fits-all nature, which produces a highly similar experience mix across military communities and occupations, despite differences in training costs and productivity and the value of experience, knowledge, and skills. Further, it induces careers that may be too short for some groups where the learning curve is long, the value of experience is high, and the physical demands are not paramount. The system is inefficient because it defers a relatively large amount of compensation over a career into the retirement benefit, despite the fact that the typical service member is young and has a preference for current over deferred compensation. Because of this preference, the military could sustain recruiting and retention while saving costs by increasing current compensation and reducing deferred compensation. Finally, the system is inequitable because the majority of service members leave before serving 20 years and thus do not qualify for retirement benefits.

That said, the current system has advantages, not the least of which is that it has generally helped the services to meet their manning requirements over the past 70 years despite tremendous changes in the military, in technology and manpower requirements, in DoD's wartime and peacetime posture, and in the civilian population and economy. The system helps provide a predictable and stable career force to the services and provides advantages to service members. The DB feature of the current system, backed by the federal government, is low-risk and predictable, and the immediate annuity to qualified personnel provides a transition benefit and continuing source of support to members as they engage in their second career in the civilian world.[1]

The MCRMC sought retirement alternatives that maintained the advantages of the current system while addressing the criticisms. Recent studies, reviews, and commissions have highlighted the advantages of a blended plan for achieving these objectives.[2] DoD's *Concepts for Modernizing Military Retirement* (2014), the *Defense Science Board Task Force on Human Resources Strategy* (Defense Science Board, 2000), the Defense Advisory Committee on Military Compensation (2006), and the Tenth QRMC (U.S. Department of Defense, 2008a and 2008b) recommended that the current military retirement system be replaced with a blended approach that includes a less generous DB plan, early vesting in a defined contribution (DC) plan with DoD contributions, and higher current compensation. The details of the blended approach—such as when to vest, how to reduce the DB plan, and how current compensation should be increased—varied across studies. The studies generally found that a blended approach increased efficiency, reduced cost, increased flexibility, and increased equity.

The MCRMC proposal for retirement reform is also a blended approach. As discussed in more detail in Chapter Two, it includes the following features for AC members:

- DB retirement program, vested at YOS 20 with immediate benefits for vested members, using a formula of 2% × YOS × average of the highest three years of basic pay. Members can choose to receive a lump sum in place of all or part of the annuity between the member's age at retirement and age 67.

[1] "Second career" has been used in the literature to describe working in the civilian economy after retiring from the military. This period is also referred to as "working age" years after the military.

[2] In several of these past studies, the blended approach was referred to as a *hybrid* approach.

- DC retirement program, vested at the beginning of YOS 3, with an automatic DoD contribution of 1 percent of basic pay starting at the beginning of the first YOS and ending at the conclusion of YOS 20. The program also has a matching element, with the DoD matching member contributions up to 5 percent of basic pay, starting at the beginning of YOS 3. The member and DoD matching contribution default to 3 percent of basic pay unless the member chooses to raise or lower his or her contribution. Members can opt-out of the matching program entirely and still receive the automatic contribution.
- Continuation pay paid to members at YOS 12, as a multiplier of monthly basic pay, to sustain retention.

To support its assessment of alternative ideas for retirement reform, the Commission sought general analytical support from RAND as well as modeling and cost analyses. RAND has a substantial body of research and analysis related to military compensation and retirement policy, including research in support of numerous previous QRMCs. Past research includes comparisons of military and civilian pay and the development, estimation, and application of a stochastic dynamic programming model, known as the DRM of active and reserve retention. The application of the DRM involves simulations of the impact of compensation and retirement policy changes on AC and RC retention, as well as on cost and outlays, in the steady state and during the transition to the steady state. RAND recently used the DRM approach to simulate the cost and retention effects of military retirement reform alternatives proposed by the internal DoD working group (Asch, Hosek, and Mattock, 2014) and summarized in the DoD white paper that was intended to inform the Commission's deliberations (U.S. Department of Defense, 2014). Other recent RAND studies using the DRM approach include analyses of reserve retirement reform, analyses of Air Force aviator retention and the effects of aviator special and incentive pays, and analyses of DoD civil service retention and the effects of pay freezes and retirement program changes. Details of the RAND DRM of military active and reserve retention and cost have been presented in past studies and are not presented here. Instead, we offer a brief summary of the model in Chapter Three.

RAND's research support involved three tasks. The first was to provide analysis of the steady-state retention effects of the MCRMC retirement reform proposal, to estimate the amount of continuation pay required to sustain retention under the reform, and to develop a formula for a lump-sum option in place of a DB annuity between retirement and age 67. The second task was to estimate the steady-state cost savings of the reform proposal, working in collaboration with the DoD Actuary. The final task was to analyze the cost savings and change in government outlays during the transition to the steady state, including permitting currently serving members to opt in to the new system. Retired member opt in to the new system was not modeled. This document summarizes the analysis we performed for each of these tasks.

To accomplish the tasks, RAND adapted the simulation capability of the DRM to incorporate the features of the MCRMC retirement reform proposal and then simulated the steady-state and transitional retention effects for each service, for officers and for enlisted personnel, and for AC and RC personnel. We also simulated the change in costs in the steady state in the transition years as well as the change in Treasury Department outlays in the transition.

Chapter Two describes the features of the retirement reform proposal and describes our approach for computing the lump sums. Chapter Three presents an overview of the DRM and its simulation methodology. Chapter Four contains our steady-state simulation results, while Chapter Five describes the results for the transition period. Chapter Six provides concluding thoughts. The main text of the documents presents illustrative examples of the results, while the report's two appendixes provide more-detailed results for all cases we analyzed.

CHAPTER TWO

Elements of the MCRMC Retirement Reform Options

The retirement reform package MCRMC asked RAND to analyze consists of three main components: the addition of a DC plan, changes to the current DB plan, and changes to current compensation, in the form of an incentive pay to sustain retention. We describe each of these elements in turn, first for the AC and then for the RC. The elements of the reform are summarized in Table 2.1 for the AC and in Table 2.2 for the RC.

The Defined Contribution Plan

The DC plan consists of two parts: an automatic contribution and a matching contribution. Members would be enrolled in the Thrift Savings Plan upon enlistment or accession, and DoD would make an automatic contribution of 1 percent of a member's basic pay starting at entry and continuing through YOS 20. Members would fully vest in the fund at the beginning of YOS 3.

DoD would match member contributions of up to 5 percent of a member's basic pay at the beginning of YOS 3. Members would automatically be enrolled in the plan to contribute 3 percent of their basic pay to the fund. Participation in the matching part of the plan is voluntary, and members could choose to increase or decrease their own contribution or opt-out entirely.

Members would have full access to the funds in their Thrift Savings Plan upon reaching age 59½.

Changes to the Defined Benefit Plan

The DB plan would change the multiplier from 2.5 percent to 2.0 percent; that is, it would change the value of the retirement annuity from 2.5% × YOS × average of the highest 3 years of basic pay, to 2.0% × YOS × average of the highest 3 years of basic pay.

Upon AC retirement, members would be offered the option to choose between (1) receiving the regular (2-percent multiplier) full annuity immediately; (2) receiving a partial lump-sum payment along with a reduced annuity based on a multiplier of 1 percent up to age 67 and receiving the regular annuity thereafter; or (3) receiving a full lump-sum payment along with the regular annuity that would begin at age 67, but there would be no annuity payments from the age of retirement to age 67. Vesting for the DB plan would continue to be upon completion of YOS 20.

Table 2.1
AC Plan Parameters

Parameter	Value
DB vesting at YOS	20
DB multiplier	2.0%
DB or lump-sum options from retirement to age 67	Yes
DB age of full retirement	67
DC retirement eligibility age	59½
DC DoD matching contributions YOS	3 to 20
DC DoD automatic contributions YOS	1 to 20
DC DoD automatic contribution	1%
DC vesting YOS of automatic DoD contribution	3
DC DoD matching contribution	3% auto, up to 5%, or 0% if opt out
DC vesting YOS of matching DoD contribution	3
DC growth rate[a]	4%
DC member contributions included in computation of fund value?	No
DC member contributions subtracted from current military earnings?	No
AC continuation pay multiplier (in months of basic pay)	Basic multiplier = 2.5, additional pay to be provided by service[b]
AC continuation pay payout structure	Lump sum at YOS 12, with pro-rated payback if 4 additional years of service not completed

[a] The 4-percent growth rate is an assumption made in modeling the proposal, at the request of the MCRMC.

[b] In the policy simulations, the AC continuation pay multiplier together with the RC continuation pay multiplier are optimized to sustain AC retention and RC participation.

The exact values of the lump-sum payments were not specified by the MCRMC, leaving this task to DoD as part of its implementation of the new system. But, for analysis purposes, the MCRMC asked RAND to determine a formula or schedule for lump-sum payments that would be based on the average of the member's highest 3 years of basic pay, YOS, and the number of years remaining until the member reaches age 67. The formulas for the AC and for the RC that we developed differ and are given in Table 2.3. We describe the development of the formula in Appendix B.

Changes to Current Compensation

Members would receive a continuation pay at YOS 12, the exact amount to be determined by the services. This pay would be associated with an obligation to continue for four more years. Members who voluntarily leave the force before their four-year obligation is completed would be required to pay back the continuation pay received to date on a prorated basis. For example,

Table 2.2
RC Plan Parameters

Plan Element	Value
DB vesting at YOS	20
DB multiplier	2.0%
DB between retirement and age 60	No
DB retirement eligibility age, or immediate lump-sum choice	60
DB full retirement age	67
DC retirement eligibility age	59½
DC DoD matching contributions YOS	3 to 20
DC DoD automatic contributions YOS	1 to 20
DC DoD automatic contribution	1%
DC vesting YOS of automatic DoD contribution	3
DC DoD matching contribution	3% auto, up to 5% or 0% if opt out
DC vesting YOS of matching DoD contribution	3
DC growth rate[a]	4%
DC member contributions included in computation of fund value?	No
DC member contributions subtracted from current military earnings?	No
RC continuation pay multiplier (in months of AC basic pay)	Basic multiplier = 0.5 with additional amount provided by service[b]
RC continuation pay payout structure	Lump sum at YOS 12, with pro-rated payback if 4 additional years of service not completed

[a] The 4-percent growth rate is an assumption made in modeling the proposal, at the request of the MCRMC.

[b] In the policy simulations, the AC continuation pay multiplier together with the RC continuation pay multiplier are optimized to sustain AC retention and RC participation.

a member who only served one year out of the four would be required to repay three-quarters of the continuation pay received at YOS 12.

In our model, continuation pay is a multiple of monthly basic pay, and AC members would be guaranteed a minimum of 2.5 months of basic pay (i.e., a 2.5 multiplier) and RC members would be guaranteed a minimum of 0.5 month of basic pay (a 0.5 multiplier). Continuation pay multipliers were determined during policy simulations as the value producing the best fit to active and reserve force size and shape, given the other elements of the reform. These 2.5 and 0.5 multipliers selected for the AC and RC, respectively, are roughly the minimum values of the optimized multipliers. The MCRMC proposal would require that the services receive automatic funding to cover the minimum continuation pay amounts, but the services would separately have to request funds to cover any additional continuation pay to supplement these minimums. The optimized multipliers are reported in the next chapter; optimized multipliers are the values consistent with maintaining AC and RC retention given the other elements of the MCRMC reform.

Reserve Component Retirement Reform

The proposed RC retirement reform is directly analogous to that in the AC. The DC plan automatic and matching amounts are scaled to the amount of basic pay RC members receive for their calendar days of service in a year, and vesting occurs at the beginning of AC+RC YOS 3. Vesting in the RC DB plan would continue to be at YOS 20 and retirement annuity payments would begin at age 60, as under the current system. However, at separation, RC members eligible for retirement could choose between receiving that standard annuity between ages 60 and 67 and thereafter; a partial lump sum paid at separation from the RC and a reduced annuity paid between ages 60 and 67, followed by the standard annuity thereafter; or a full lump sum paid at separation from the RC and no annuity between ages 60 and 67, with the standard annuity thereafter. Those who chose a full or partial lump sum would receive the lump sum at separation and would not be required to wait until age 60, though those who opted for the standard annuity and no lump sum would be required to wait until age 60 to receive the annuity. Thus, separating RC members eligible for RC retirement benefits could receive a lump sum at separation and either no annuity, a partial annuity, or a full annuity between ages 60 and 67. Members of the RC would also receive a continuation pay beginning at AC+RC YOS 12, where we model continuation pay as a multiple of monthly basic pay at YOS 12. The parameters of the RC plan are summarized in Table 2.2.

Lump-Sum Formulas

As mentioned earlier in the chapter, AC members would have the choice between an annuity from retirement age to age 67, a partial annuity together with a partial lump sum during this period, or no annuity and a full lump sum. The values of the lump-sum payments would be determined by a formula or schedule that uses the average of the member's highest 3 years of basic pay, YOS, and the number of years remaining until the member reaches age 67. The formula for the lump sums for AC personnel is

$$\text{AC Lump Sum} = \text{M} \times \text{High-3 BP} \times \text{YOS},$$

where

M = $a - b$ age; if separation age < 55
M = $a - b$ age – 2 b (Age – 54); if separation age is ≥ 55
High-3 BP = average annualized highest 36 months of monthly basic pay
YOS = year of service.

As discussed earlier in this chapter, separating RC members eligible for RC retirement benefits could receive a full lump sum at separation from the RC and no annuity from ages 60 to 67, a partial lump sum at separation from the RC and a partial annuity from ages 60 to 67, or a full annuity between ages 60 and 67. The formula for the lump sums for RC personnel is

$$\text{RC Lump Sum} = \text{M} \times \text{High-3 BP} \times \text{YOS},$$

where

M = Maximum [$a - b$ (Years to age 60), one month of High-3 BP]; if separation age < 60
M = Sum of RC annuity payments between separation age and 67; if separation age ≥ 60
High-3 BP = average annualized highest 36 months of monthly basic pay
YOS = year of service.

The parameters *a* and *b* for the AC and RC are given in Table 2.3. These parameter values were selected by the MCRMC and informed by analysis that involved optimizing the values *a* and *b* (together with the continuation pay multipliers) to sustain retention. We describe the development of the formulas in Appendix B.

Opt in

A final feature of the proposal concerns the transition to the new plan. All currently serving members and retirees are grandfathered under the existing system, but would have the choice to opt in to the new system during an open enrollment period. We model the opt-in behavior of currently serving members but, as mentioned, we do not model the opt-in behavior of retirees. All new members who enter after the policy change would be automatically enrolled in the new system.

Table 2.3
AC and RC Lump-Sum Formula Parameters

	AC		RC	
	a	*b*	*a*	*b*
Partial lump sum	.16	.0015	.048	.002
Full lump sum	.32	.003	.96	.004

CHAPTER THREE

Brief Overview of the Dynamic Retention Model

RAND's DRM is well suited to the analysis of structural changes in military compensation, such as the MCRMC's proposed changes to the retirement system. Recent applications of the model include analyses for the ninth, tenth, and 11th QRMCs, as well as analysis in support of the recent DoD review of military compensation reform. The model's capability has steadily increased; for instance, new, faster estimation and simulation programs have been written, costing has been refined, and the model can now show retention and cost effects in both the steady state and the year-by-year transition to the steady state.

The approach is documented in several RAND reports (for instance, Mattock, Hosek, and Asch, 2012, a technical report prepared for the 11th QRMC; and Asch, Hosek, and Mattock 2013).[1]

The model is based on a mathematical model of individual decisionmaking over the life cycle in a world with uncertainty and where members have heterogeneous preferences (tastes) for active and for reserve service. The parameters of this model are empirically estimated with data on military careers drawn from administrative data files. The model begins with service in the AC, and individuals make a stay/leave decision in each year. Those who leave the AC take a civilian job and, at the same time, choose whether to participate in the RC. The decision of whether to participate in the RC is made in each year, and the individual can move into or out of the RC from year to year. More specifically, a reservist can choose to remain in the RC or to leave it to be a civilian, and a civilian can choose to enter the RC or remain a civilian. In the model, each service has a single reserve component. For example, the Army National Guard and U.S. Army Reserve are not treated separately but are combined into a single group, the Army RC, and similarly for the Air National Guard and Air Force Reserve.

Among the parameters we estimate is the personal discount factor. The discount factor indicates how much a member values a dollar today versus one year hence. Because we estimate a separate model for each service, for officers and for enlisted, we have estimated personal discount factors for each service, officer and enlisted. The estimated real personal discount factors range from 0.88 to 0.90 for enlisted personnel across the four services. That is, a dollar next year is worth 88 to 90 cents today. For officers, estimates are remarkably similar across services,

[1] Goldberg (2002) provides an extensive discussion of the history of retention models. Gotz (1990) provides a detailed discussion of the advantages of the DRM approach relative to other approaches that have been used to assess the effects of compensation reform proposals on retention.

at 0.94. These estimates, together with the other model estimates, are discussed in past documents, such as in Asch, Hosek, Mattock, 2014, Appendix E.[2]

The data we use are from the Defense Manpower Data Center Work Experience File (WEX). The WEX contains person-specific longitudinal records of AC and RC service members. We use the WEX data for service members who began their military service in 1990 or 1991 and track their individual careers in the AC and, if they join, the RC, through 2010, providing 21 years of data on 1990 entrants and 20 years on 1991 entrants. For each AC component, we drew samples of 25,000 individuals who entered the component in fiscal years (FYs) 1990 or 1991, constructed each service member's history of AC and RC participation, and used these records in estimating the model. We supplement these data with information on active, reserve, and civilian pay. AC pay, RC pay, and civilian pay are averages based on the individual's years of AC, RC, and total experience, respectively. We use 2007 military pay tables, but, because military pay tables have been fairly stable over time, with few changes to their structure,[3] we do not expect our results to be sensitive to the choice of year. For civilian pay opportunities for enlisted personnel, we used the 2007 median wage for full-time male workers with associate's degrees. For officers, we used the 2007 80th percentile wage for full-time male master's degree holders in management occupations. All data on civilian pay opportunities are from the U.S. Census Bureau.

We estimated the model separately for officers and enlisted in each of the four services. The estimation methodology and model estimates are reported in Mattock, Hosek, and Asch (2012); Asch, Hosek, and Mattock (2013); and Asch, Hosek, and Mattock (2014). We also developed simulation code that allowed us to simulate retention over the military career in both the AC and RC and to compute the cumulative retention profile in the steady state. By *steady state*, we mean when all members have spent an entire career under the reformed system. We simulate the retention profile under the current compensation system, which we call the baseline force. We then simulate retention under hypothetical alternative compensation systems, such as the MCRMC reform proposal.

Another feature of our simulation capability is the computation of personnel costs. Our cost estimates include the cost of current compensation, including basic pay, the housing allowance and the subsistence allowance, and the cost of retired pay for the steady state, baseline force.

Simulations are done for the steady state for the AC and RC and for the transition to the steady state for the AC. The transition analysis allows us to address questions about how a new system will affect members currently in service versus new members who are automatically enrolled in the new system. In our transition analysis, we model the choice of existing members to opt in to the new system. We assume AC members have a one-year open enrollment window and choose to opt in if the value of staying in the AC at the time of the choice is higher under the new system than under the existing system.[4] This allows us to compute

[2] The MCRMC final report cites personal discount rates from RAND's analyses (MCRMC, 2015, pp. 33–34). The rates cited are arithmetic means of the implied rates from the personal discount rates that we estimate.

[3] An exception was the structural adjustment to the basic pay table in FY 2000 that gave larger increases to midcareer personnel who had reached their pay grades relatively quickly (after fewer years of service). A second exception was the expansion of the basic allowance for housing, which increased in real value between FY 2000 and FY 2005. It should be noted that the costing analysis is in 2016 dollars.

[4] The MCRMC recommended a six-month window.

the percentage of members who opt in. In our transition modeling, we compute the change in cost and in government outlays in the transition to the steady state. We also can compute the change in retention during the transition, but because the reform proposal is designed to sustain retention, retention is by and large constant in the transition to the steady state as well as in the steady state. Thus, we only briefly discuss the change in retention during the transition and focus the discussion on the change in cost and outlays.

Limitations

The DRM has several limitations. The model assumes that military pay, promotion policy, and civilian pay are time-stationary, and it excludes demographic factors such as gender, marital status, and spouse employment. It also excludes health status and health care benefits, and we do not explicitly model deployment or deployment-related pay. That said, the estimated models fit the observed data well for the both the AC and the RC.

Some limitations are specific to simulating the MCRMC retirement reform proposal. The DRM does not model members' choices regarding an annuity for an AC member between age of retirement and age 67, or an annuity between ages 60 and 67 for an RC member versus a lump sum payment at separation. The DRM also does not model members' savings decisions and therefore their decisions regarding whether and how much to contribute to the DC plan. Therefore, we are not able to simulate what percentage of members will choose an annuity versus a partial lump sum versus a full lump sum, nor are we able to simulate the distribution of contribution rates among service members to the DC plan, and therefore the DoD DC match rate.[5]

Given these limitations regarding the ability of the DRM to model the MCRMC proposal, our approach involves simulating the retention effect of the MCRMC reform under alternative assumptions about the lump-sum choice and the DoD DC match rate. That is, we conduct simulations under the assumption that all members make a given lump-sum choice and all members receive a given DoD DC match rate. Given the three lump-sum possibilities (all members receive no lump sum, all receive a partial lump sum, all receive a full lump sum)[6] and three assumptions about DoD matching (all members receive a 0-percent match, all receive a 3-percent match, and all receive a 5-percent match), there are nine alternatives of the reform that we considered.[7]

Having a choice of a lump sum or annuity is a valuable feature of the reform package. Similarly, the availability of a DoD matching contribution is a valuable feature. This additional value also improves the value of staying in the military and therefore improves reten-

[5] The DRM could be extended to include these decisions. However, data are not available to allow an empirical implementation of this extension.

[6] In fact, there are more than three possibilities because there are three possibilities for the AC lump-sum choice and then another three possibilities for the RC lump-sum choice, or a total of nine for any given DC match rate. Given the three DC match rate cases, there are 27 possible cases in total for a given service and rank (officer or enlisted), or a total of 27 × 4 (services) × 2 (ranks) = 216. To cut down on the number of simulations, we considered only three rather than nine lump sum choices: (1) AC chooses no lump sum/RC chooses no lump sum; (2) AC chooses partial lump sum/RC chooses partial lump sum; and (3) AC chooses full lump sum/RC chooses full lump sum.

[7] In all cases, the member receives the automatic 1-percent DoD contribution; there is no obligation for the member to match this contribution.

tion. Because we do not model the lump sum choice or DC choice, we are unable to include the value of having these choices as an element of the overall value of staying in the military. Thus, we understate the retention effect of the reform package by an unknown amount. The overall retention effect in our simulations is a weighted average of retention under the nine alternatives, described above, not accounting for the unknown retention effect of the value of choice regarding the lump sum or DC matching contribution. In implementation, we put zero weights on some of the nine and fractional weights on others, to generate the low cost savings case, the high costs savings case, and the middle cost savings example. Still, since the main result we find and describe in the next chapter is that retention under each alternative is virtually identical to the baseline, the weighted average will also be virtually identical. The optimized continuation pay multipliers that sustain the force would be lower, and their cost lower, if we could incorporate the value of choice.

In presenting the retention results, we show results in the main text for the cases when all members receive a 3-percent DC match rate and all members choose either an annuity, a partial lump sum/partial annuity, or a lump sum and no annuity (for AC members between retirement age and age 67, and for RC members between ages 60 and 67). Also, in the main text we generally show the results only for the Army. The results for the other cases and for the other services are in Appendix A.

In presenting the cost and outlay results, we focus on three cases (as discussed in more detail in the next chapter); namely, the minimum cost savings case, the maximum cost savings case, and an example of a middle cost savings case. The minimum cost savings case corresponds to the case where all members receive an annuity and a 5-percent DC match rate. In the maximum cost savings case, members receive a 0-percent DC match rate and receive a full lump sum with no annuity. The example of a middle case is where officers and enlisted personnel receive a 3-percent DC match rate and we assume 50 percent of enlisted members choose the annuity, 25 percent choose a partial annuity/partial lump sum, and 25 percent choose a full lump sum with no annuity, while 100 percent of officers choose the annuity. None of these cases is realistic, because we expect some variation in the DC match rate across personnel and variation in their annuity versus lump sum choices. Still, we show the minimum and maximum cases to illustrate the range of effects. We show the middle case as an example of what a mid-point case between the minimum and maximum cases might be, but obviously other examples are possible.

Finally, there is the notion of cost savings itself. By *cost savings*, we mean the extent to which costs are lower relative to the costs of the current system. The costs of the current system are reflected in DoD's military personnel budget, which includes the retirement accrual charge. In this accounting, a decrease in the accrual charge under the MCRMC proposal relative to the current system registers as a cost savings. However, it should be kept in mind that Congress could choose to reduce DoD's top line by that amount, so the "savings" would not be available to allocate to other purposes. Alternatively, the top line might be left intact, in which case DoD would purchase more of something—for example, more operations and maintenance, or more procurement. Further, because the accrual charge is an inter-governmental transfer from DoD to the Treasury, it does not entail any outlays. This implies that if Congress did decide to reduce DoD's top line by the amount of the decrease in the accrual charge, this would produce no decrease in federal outlays and, hence, would have no effect on the current deficit.

Optimization

An important objective of the current analysis is for the MCRMC reform proposal to support a service's current force size and structure, even if they might be changed in the future. To achieve this in our model, our simulations compute optimized values of continuation pay and lump sums, given the other features of the reform proposal, such as the retirement benefit formulas and vesting and eligibility conditions. The optimization aspect of the simulation capability helped tune the MCRMC proposal to meet manning objectives and provided information needed to support the selection of the continuation pay multipliers of 2.5 for AC and 0.5 for RC.

As discussed in the previous chapter, continuation pay for the AC and for the RC equals a continuation pay multiplier times the active duty monthly basic pay at YOS 12, with a payback feature for those who separate before completing four additional years of service. As discussed in Appendix B, the lump-sum payment is based on a formula with parameters *a* and *b* for the AC and for the RC.

Determination of the values of the continuation pay multipliers for AC and RC and the parameters for the AC and RC lump sums was a two-step process. The first step involved computing the continuation pay multipliers and the lump-sum parameters that optimize the fit to the current AC and RC force size and shape, given the DB and DC policy elements. First, we optimize the continuation pay multipliers, assuming all members choose an annuity rather than either of the lump-sum options; that is, we first optimize the continuation pay multipliers, assuming the annuity choice. Second, we optimize the lump-sum parameters, given the optimized continuation pay multipliers, assuming all members choose the partial lump-sum option, and again assuming that all members choose the full lump-sum option.

The second step involved the selection of fixed lump-sum parameters for the AC and for the RC. The MCRMC required that these parameters be fixed across enlisted personnel and officers and across all services for the AC and for the RC. Fixing these parameters permits the lump sum to be funded on an accrual basis and be part of the military retirement fund. The MCRMC selected the fixed lump-sum parameters using the optimized values from the first step as a guide. Table 2.3 in the previous chapter shows the selected parameter values while Chapter Four and Appendix B discuss the optimized values of the continuation pay and lump-sum parameters.

The simulation results we present, as well as the optimized continuation pay multipliers we show in Chapter Four and Appendix A, use the continuation pay multipliers optimized in the first step together with the fixed lump-sum parameters in Table 2.3 that were informed by the optimization of the lump-sum parameters, described in Appendix B.

CHAPTER FOUR

Steady-State Retention Results

In this chapter, we present AC retention and RC participation results for each variant of the MCRMC reform proposal, by service, for officers and for enlisted personnel. By *variant*, we mean the different combinations of lump-sum choices and DoD DC rates. Because the results are qualitatively similar across the services for AC retention and RC participation for officers and for enlisted personnel, we focus here on the results for the Army and have placed the results for the Navy, Marine Corps, and Air Force in Appendix A. We then discuss the optimized continuation pay and lump-sum parameters that sustain retention under the MCRMC reform. The last part of the chapter presents our steady-state estimates of cost savings for three cases, making use of input received from the DoD Actuary, for the AC and the RC. The next chapter presents estimates of the changes in cost and outlays from the current system for the transition period.

AC Retention and RC Participation

Figures 4.1a–4.1d show AC retention and RC participation results for Army enlisted and officer personnel. The simulations in these figures use the values of the continuation pay multipliers conditional on the values of the lump-sum multiplier parameters for AC and RC in Table 2.3. That is, the simulation presets the lump-sum parameters using the values in Table 2.3, and, as requested by MCRMC, these values are the same for officers and enlisted personnel. Then, given these values, the simulation optimizes the continuation pay multipliers. As discussed in Chapter Three, by *optimized*, we mean the values that minimize the difference between retention rates under the reform alternative and the baseline. Also, as we discuss later in this chapter and in Appendix B, the lump-sum parameters in Table 2.3 generally correspond the values that would result if the lump-sum parameters were optimized along with the continuation pay multipliers for enlisted personnel. But the lump-sum parameters in Table 2.3 tend to be too low for officers; that is, they are less than the values that would result if they were optimized along with the continuation pay multipliers. This is a consequence of the DRM's estimated personal discount rates for officers being less than those of enlisted personnel. As a result, the lump sum is comparatively low for officers, and most officers will find it more advantageous to select the annuity option over the lump-sum options. Thus, we show only the AC retention and RC participation results for the annuity case for officers, but show the results for the annuity and lump-sum choices for enlisted personnel.

The baseline in the figures corresponds to the retention and participation outcomes under the current compensation system. While there was no presumption that future requirements will call for the same size and mix as the baseline force, a criterion for assessment is that

Figure 4.1a
Army AC Enlisted Retention and Prior-Active RC Enlisted Participation, 3-Percent DoD Match Rate, Annuity-Only Choice

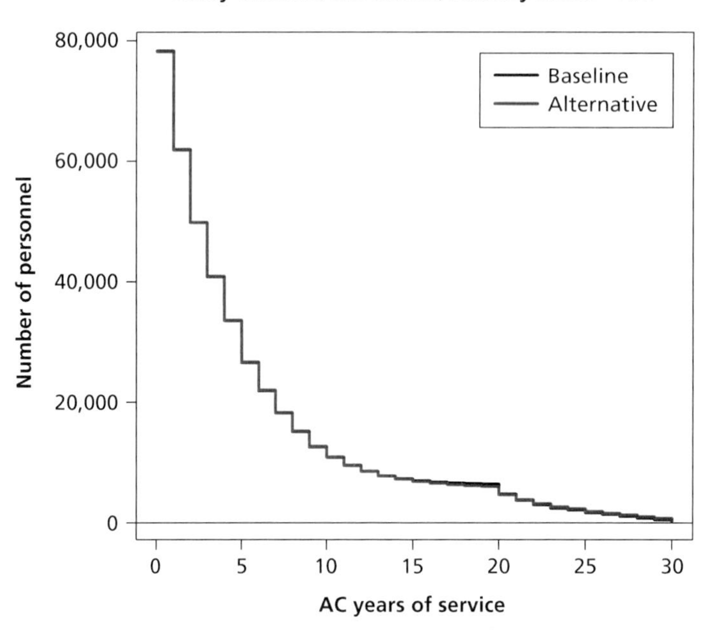

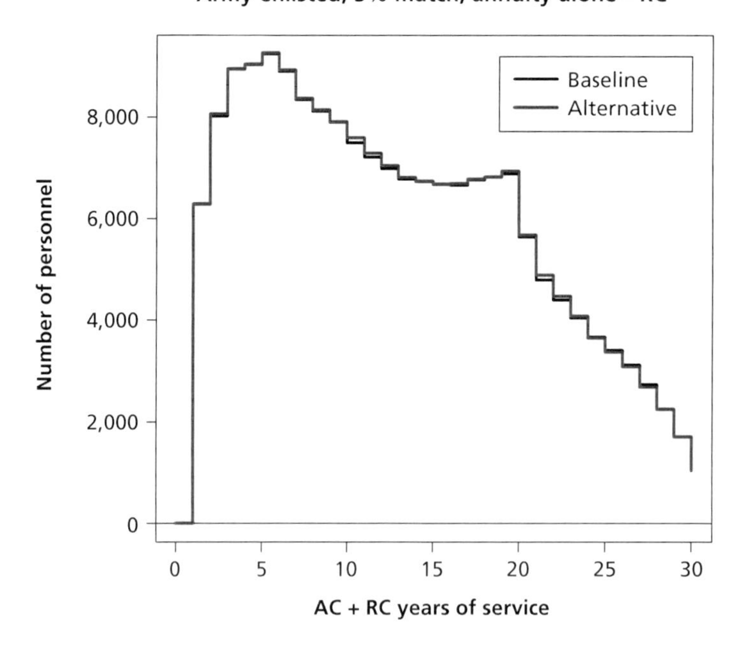

RAND *RR1022-4.1a*

Figure 4.1b
Army AC Officer Retention and Prior-Active RC Officer Participation, 3-Percent DoD Match Rate, Annuity-Only Choice

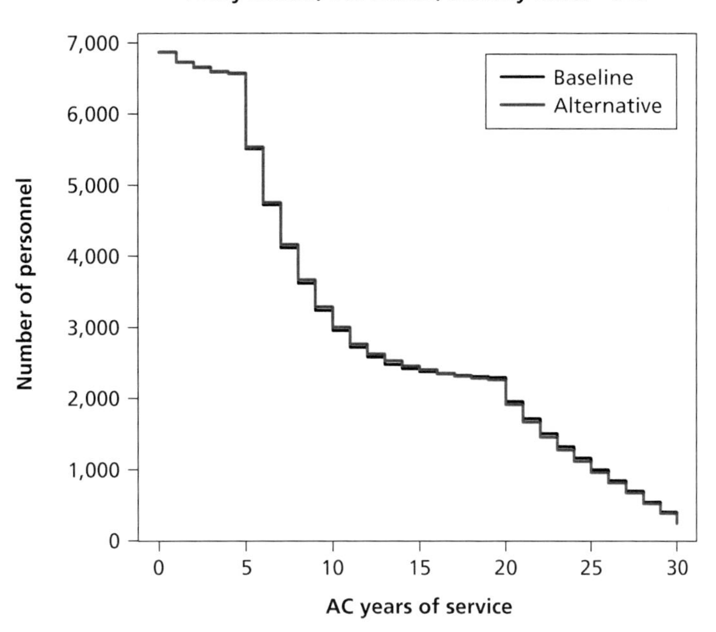

Army officer, 3% match, annuity alone—RC

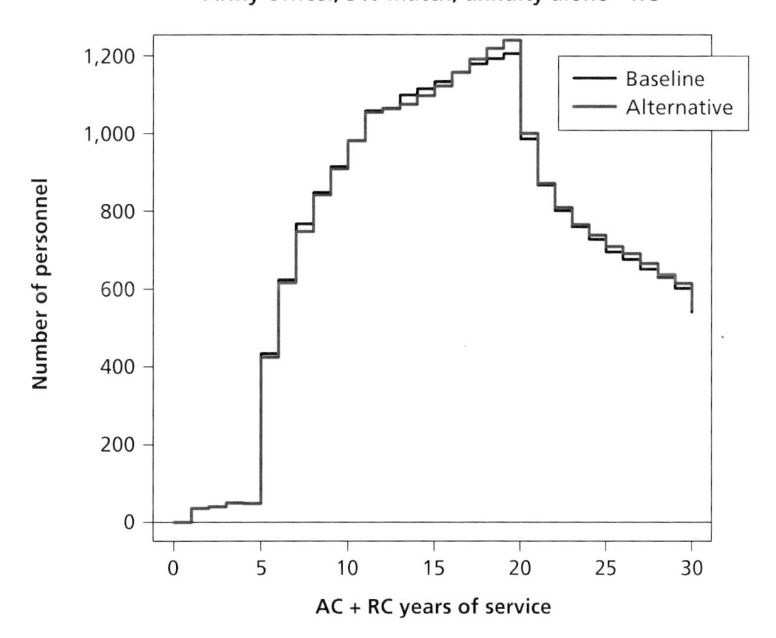

RAND *RR1022-4.1b*

Figure 4.1c
Army AC Enlisted Retention and Prior-Active RC Enlisted Participation, 3-Percent DoD Match Rate, Partial Annuity/ Partial Lump-Sum Choice

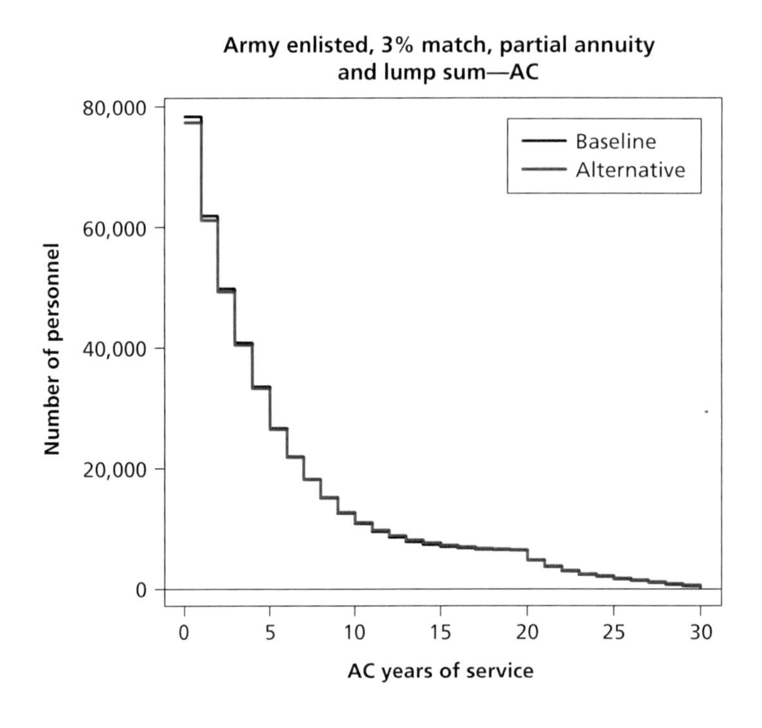

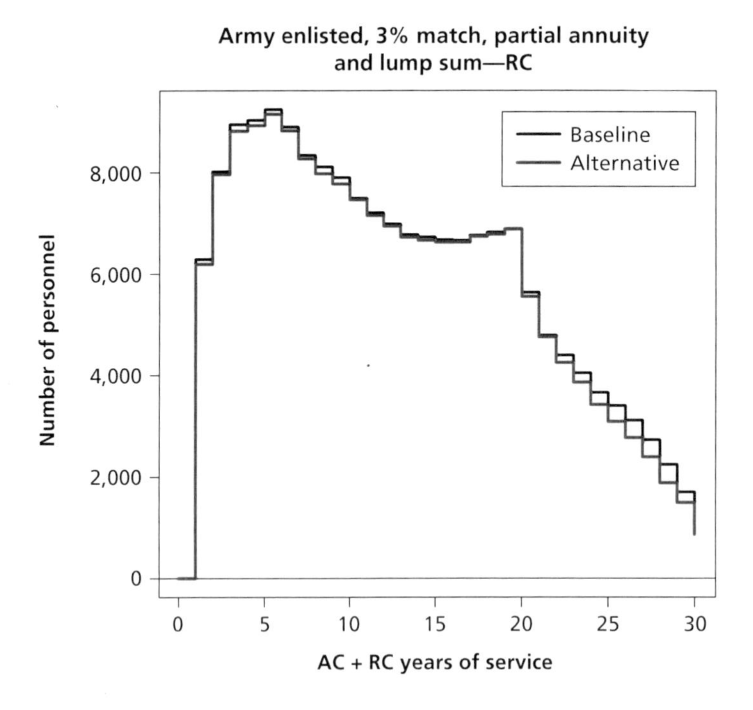

RAND *RR1022-4.1c*

Figure 4.1d
Army AC Enlisted Retention and Prior-Active RC Enlisted Participation, 5-Percent DoD Match Rate, No Annuity/Full Lump-Sum Choice

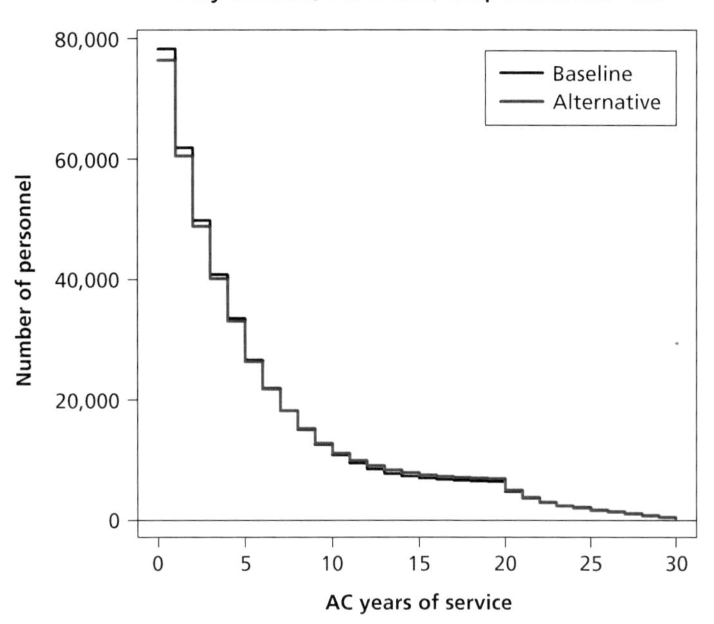

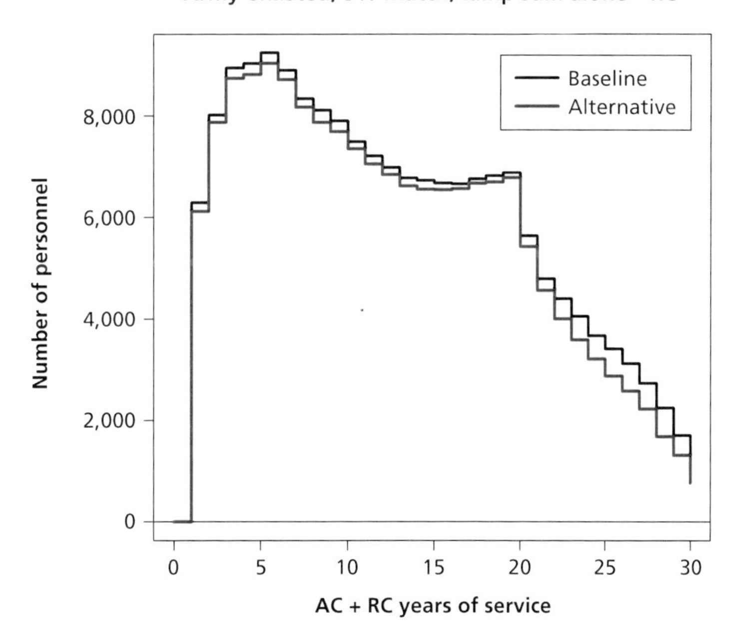

RAND *RR1022-4.1d*

any reform alternative should be able to achieve the baseline. Furthermore, comparing cost-effectiveness of alternatives requires that effectiveness, namely the ability of the alternative to achieve a given force profile, be held constant.

Figure 4.1 shows results for Army enlisted and officer personnel, assuming 3-percent DoD matching, so that members receive the 1-percent automatic DoD contribution plus the default 3-percent DoD matching contribution. Figures 4.1a and 4.1b show AC and RC results, for enlisted personnel and officers, respectively, assuming all members choose an annuity between retirement and age 67 for AC retirees and for ages 60 to 67 for RC retirees. Figure 4.1c shows similar results for enlisted personnel but assumes AC enlisted retirees choose a partial annuity/partial lump sum between retirement and age 67 phase and RC enlisted retirees choose a partial annuity/partial lump sum from ages 60 to 67. Similarly, Figure 4.1d shows the full lump-sum case (and no annuity between retirement and age 67) for AC and for RC enlisted personnel. In the figures, Army AC results are the upper graphic and RC results are below. Figures A.1 and A.2 in Appendix A have a similar format to Figure 4.1, but Figure A.1 shows results assuming a 0-percent DoD match rate to the DC plan, implying that members receive only the automatic 1-percent DoD contribution. Figure A.2 shows results assuming a 5-percent match rate, for a total contribution rate of 6 percent.[1]

The main finding is that AC retention and RC participation are quite close to baseline retention and participation regardless of the DoD matching rate and regardless of the lump-sum option choice. Since continuation pay and the lump-sum choice are elements of a package of reforms working simultaneously—the DC plan, the DB annuity, continuation pay, and the lump-sum option—it is more accurate to say that, given the parameters of the DC plan and DB annuity (multipliers, vesting, age of benefit receipt) and the parameters of continuation pay and lump sums (formula, when paid), the optimized multipliers for continuation pay and the lump sums perform well to sustain AC retention and RC participation. The package is effective.[2]

These simulations show that the addition of a DC plan that vests at YOS 3 and permits payout as early as age 59½, together with continuation pay at YOS 12, are sufficient to sustain retention and participation despite the lower DB multiplier under the reform—2.0 percent rather than 2.5 percent. If there were no DC plan and no continuation pay, the lower DB multiplier would result in fewer mid-career personnel staying until YOS 20 and fewer would be induced to leave after YOS 20. But when the DC plan and continuation pay are included, the DoD contributions to the DC plan from YOS 3 through YOS 20, together with the continuation pay at

[1] In sum, Figures 4.1a, A.1a, and A.2a show AC enlisted retention and prior-active-service RC participation with annuity choice for the Army; Figures 4.1b, A.1b, and A.2b show the Army officer results with the annuity choice; Figures 4.1c, A.1c, and A.2c show Army AC enlisted retention and prior-active-service RC participation with the partial annuity/lump-sum choice; and Figures 4.1d, A.1d, and A.2d show AC enlisted retention and prior-active-service RC participation with the no annuity/full lump-sum choice. The format of the figures is similar for the other services, shown in Appendix A. These other cases are in the Appendix to cut down on the number of graphics in the main text for ease of readability, not because they are less important.

[2] The fit for the Army enlisted RC among those with prior AC service is not quite as good for the lump-sum choices in Figures 4.1c and 4.1d as for the annuity choice in Figure 4.1a. This could be in part because the parameters defining the lump sums are fixed, to enable the cost of the lump sums to be included in the accrual charge, and in part because of our extreme assumption that all members take a lump-sum option. Earlier results (not shown), where we do not fix the lump-sum parameters but optimize them, show a better fit relative to the baseline for the RC. Insofar as RC participation falls short, the services may need additional RC special and incentive pays to sustain retention. Still, the good fit for the annuity case gives us assurance that RC participation can be sustained under the MCRMC proposal.

YOS 12, induce mid-career personnel to stay until YOS 20 at the same rate as the baseline and induce those reaching YOS 20 to leave at the same rate, despite the lower DB multiplier.

In the analysis, we found some trade-off between the optimized continuation pay levels and the horizon over which DoD makes DC plan contributions (not shown). If DoD made DC plan contributions for those with more than 20 years of service, e.g., contributions from YOS 3 through separation, the optimized continuation pay multipliers would be lower. That is, the amount needed to sustain retention through continuation pay is lower when DoD extends the horizon over which contributions are made. However, preliminary cost-savings estimates indicate that savings are somewhat lower, suggesting that continuation pay is a more cost-effective means of sustaining pre-YOS 20 retention and that some of the contribution to those beyond YOS 20 is "rent" (a payment in excess of that necessary to sustain retention), meaning that members who stay after YOS 20 would have likely stayed without the contributions after YOS 20.

The figures also show that the reform sustains retention and participation relative to the baseline under a range of DoD matching rates, from the lowest level of 0 percent up to the highest level of 5 percent. As one would expect, the optimized continuation pays for AC and RC are lower when the DC matching rate is higher. The optimized multipliers are discussed further in the next section.

The excellent fits relative to the baseline also indicate that the linear approximation we use and the lump-sum parameter values set by the MCRMC work well for enlisted personnel, particularly for the AC. MCRMC made its decision on parameter values based on preliminary results showing the optimized values for the lump-sum parameters; the optimized values differed between enlisted and officers and, to some extent, across the services, and MCRMC set values that were closer to the enlisted optimized values. Although the lump sum was relatively small for officers and, as mentioned, the results in the figures assume officers chose the annuity, officer continuation pay was nevertheless sufficient to sustain officer retention.

In sum, the reform performs well in being able to reproduce AC retention and RC prior-AC service participation profiles, regardless of the lump-sum option chosen or the DoD match rate. This is a valuable finding in view of the fact that data do not exist to allow estimation of the member's choice among the alternatives of full annuity, partial annuity/partial lump sum, and full lump sum. Our simulation results imply that the objective of sustaining baseline retention is robust to whichever alternative is chosen.

Continuation Pay and Lump-Sum Multipliers

Table 4.1 contains the optimized AC and RC continuation pay multipliers for each alternative, for enlisted personnel and officers, by service. As discussed in Chapter Three, multiplier values were determined in each policy simulation as the values optimizing the fit to the current AC and RC force size and shape. In short, given the decrease in the DB multiplier from 2.5 percent to 2.0 percent, the DC plan contributions (given vesting and YOS range of contributions) and continuation pay (based on the optimized multipliers) are the amounts needed to sustain retention relative to the baseline. If the reform were implemented, the services would not necessarily be constrained to apply the same continuation pay multiplier but could vary it across occupations and skill/experience levels depending on retention targets. The DRM used in our analysis of the MCRMC proposal does not distinguish between occupations and considers retention behavior for all personnel in the officer or enlisted force for each service. Thus, the optimized figures are the ones that, at the margin, provide sufficient incentives across the force to sustain the overall size of the force.

As discussed in Chapter Two and Appendix B, the lump-sum payments would be based on a multiplier, the average of the three highest years of annual basic pay, and years of service. For the AC, the formula for the multiplier for those under age 55 is $a - b \times$ (Age at Retirement) and is $a - b \times$ (Age at Retirement) $- 2 \times b \times$ (Age at Retirement $- 54$) for those ages 55 and older. The multiplier is a fraction by which the average of the three highest years of basic pay times YOS is multiplied in computing the lump sum. The multiplier formula is different for those retiring from the military after age 55 because of nonlinearity in the lump sum with respect to age, as explained in Appendix B. The lump-sum parameters a and b were obtained in two steps. First, in policy simulations we optimized the values of these parameters by service for officers and enlisted and, second, MCRMC deliberated on the results and chose parameter values that would apply across the services and to officers and enlisted alike.

Table B.1 in Appendix B contains the optimized AC and RC lump-sum parameters, assuming all members choose the full lump sum and no annuity between retirement and age 67 for those in the AC and between ages 60 and 67 in the RC. Similarly, Table B.2 shows the optimized partial lump-sum parameters, assuming all members choose the partial annuity in the AC between retirement and age 67 and a partial annuity between ages 60 and 67 in the RC. The values in Tables B.1 and B.2 are optimized to bring AC and RC retention as close to baseline as possible, given the DB and DC plans and the optimized continuation pay multipliers. On the basis of these optimized lump-sum parameters, the MCRMC selected the fixed parameters in Table 2.3 that would be the same for all members, regardless of the DC match rate. We then used these values in another round of simulations, in which we optimized the continuation pay multipliers conditional on the chosen values of the lump-sum parameters.

AC and RC continuation pay is paid at YOS 12 for enlisted members and officers. The amount equals the continuation pay multiplier times active-duty monthly basic pay at YOS 12, with one-half paid upon reaching YOS 12 and the remainder paid on three anniversaries. This approach is also used for RC continuation pay. As discussed in Chapter Two, the lump-sum payment, available instead of the partial or full annuity from retirement to age 67, is equal to the average of the highest three years of annual basic pay times YOS times a multiplier. For the RC, the lump sum is in lieu of the annuity between ages 60 and 67 and is available at separation. In the RC case, annual basic pay in the lump-sum formula is the average of the highest three years of RC basic pay and YOS is based on RC retirement points, as is currently the case.[3]

The optimized AC enlisted continuation pay multipliers range from 2.2 for the Air Force, given a 5-percent DoD DC match rate, to 5.6 for the Navy, given a 0-percent DoD DC match rate. Monthly basic pay for an E-5 with 10–12 years of service was $3,076 in FY 2014, so continuation pay would range from about $6,700 to about $17,200. As mentioned, the multipliers are lower when the DoD contribution rate to the DC plan is higher. The higher DC plan contribution rate increases the expected value of the member's DC fund and has a positive retention effect. Consequently, the continuation pay multiplier required to sustain retention is lower, resulting in a lower optimized value.

[3] Recall that the DRM analysis does not include deployment. For the RC, this means that activation and deployment are not considered when determining the highest three years of annual RC basic pay. Stated differently, the policy followed in the simulation is to use the highest annual RC basic pay for a "standard" RC year, without activation or deployment.

The optimized continuation pay multipliers for officers follow the same pattern as for enlisted but are larger across the board. For the AC, they range from about 14 to 18 months of monthly basic pay for the 0-percent DoD matching contribution rate, to about 10 to 14 months of basic pay for the 5-percent matching rate. Monthly basic pay for a major with 10–12 years of service is $6,593 in FY 2014, so the optimized continuation pay would range from about $66,000 to $119,000.

In actuality, not all members will receive a given DC matching contribution rate. Members will vary in their contribution rates, so the matching rate will also vary. Thus, the values shown in Table 4.1 should be considered as bounds for the high and low range of continuation pays required to sustain the enlisted and officer forces.

Also, the MCRMC proposal sets a basic continuation pay multiplier of 2.5 for AC members and 0.5 for RC members. Insofar as the optimized continuation pay multipliers in Table 4.1 exceed these values, the services would need to provide additional continuation pay to sustain the force. Similarly, insofar as the optimized multipliers are less than these values, the services will need to restrict retention if it is greater than required.

Enlisted multipliers are lower in the RC than in the AC. This is unsurprising since the continuation pay formula in the RC is based on the AC monthly basic pay amount. Given that RC retirement benefits are typically lower than AC benefits at a given YOS, because RC members are part-timers, a smaller fraction of AC basic pay is required to sustain retention. RC multipliers range from 0.4, for Air Force enlisted personnel when the DoD matching rate is 5 percent, to 1.6, for Marine Corps enlisted personnel when the rate is 0 percent, or a range of about half a month to two months of AC monthly basic pay. As with the AC, optimized RC continuation pay multipliers are lower when the DC matching rate is higher. Officer continuation pay multipliers follow the same pattern but are larger and range from about 5 months of basic pay to about 8 months, depending on the matching rate.

Table 4.1
Optimized Continuation Pay Multipliers for AC and RC, by Service

	DoD DC Match: 0%		DoD DC Match: 3%		DoD DC Match: 5%	
	AC	RC	AC	RC	AC	RC
Enlisted						
Army	3.1	0.9	2.8	0.9	2.5	0.5
Navy	5.6	1.5	4.8	1.2	4.2	0.8
Air Force	2.8	0.8	2.4	0.8	2.2	0.4
Marine Corps	5.0	1.6	4.2	1.1	3.7	0.9
Officer						
Army	15.2	7.1	13.0	6.2	11.6	5.8
Navy	17.9	7.5	15.2	6.7	13.5	6.5
Air Force	18.4	7.2	15.9	6.4	14.4	6.4
Marine Corps	13.5	6.2	11.7	5.8	10.3	5.1

Steady-State Cost-Savings Estimates

Though the MCRMC retirement proposal was designed to modernize the retirement system, it was important to determine whether the proposal would cost more, or less, than the current system. The MCRMC, therefore, requested that we provide estimates of the change in cost, and for this we worked with the DoD Actuary. The Actuary used its actuarial projection model, GORGO, to provide cost estimates apart from continuation pay costs, which we provided. The Actuary routinely uses GORGO when costing the military retirement system.[4]

One possible source of cost savings derives from the lower multiplier in the DB retirement formula, a change from the current 2.5 percent to the lower 2.0 percent. However, the MCRMC proposal also includes contributions by DoD to the DC plan, representing a new source of cost. In addition, to sustain retention, the proposal also included continuation pay, paid out at YOS 12. The movement of compensation from deferred compensation, in the form of a reduced DB multiplier, toward current compensation, in the form of continuation pay, over and above special and incentive pay that military personnel already receive, is a source of cost savings. Because it costs more to pay future deferred benefits than those benefits are worth to service members, given their personal discount rates, moving pay from the future to the present can save cost and still sustain retention. The cost of future benefits is greater than the value to members because they have personal discount rates that are higher than the government discount rate. That is, members discount future benefits at a higher rate than the government discounts the cost of those benefits. In practice and as supported by the DRM parameter estimates of the personal discount rate, a dollar tomorrow is worth about 94 cents today to an officer and 90 cents to an enlisted member, but the government must invest 97 cents today to pay tomorrow's dollar. Bringing compensation forward creates gains from trade. An officer or enlisted member would be better off with a payment of 95 cents today than a dollar tomorrow, and it would cost the government 2 cents less to pay the 95 cents today. The amount of the cost savings depends on the how much the cost associated with retired pay decreases and how much pay is moved forward into current compensation.

The availability of a lump-sum option is also a potential source of cost savings, insofar as members choose a lump-sum option. A lump sum, paid at separation, is less costly to provide than an annuity, because the lump sum represents the movement of deferred compensation in the form of the annuity toward higher current compensation. Thus, cost savings will be larger when more members take a lump-sum option, and larger when members take a full lump sum rather than a partial lump sum.

For our analysis of costs, the DoD Actuary prepared estimates of the annual steady-state retirement cost and cost savings for the MCRMC proposal. The estimates assumed that the MCRMC proposal could, with some force management effort by the services, maintain the baseline force size and experience mix. Our simulations showed little difference between the steady-state force under the proposal versus the baseline, so the working assumption was that any differences were small enough to handle through personnel policy actions at little cost. As a result, the Actuary's estimates focus on the accrual cost of the baseline force.[5] The Actuary

[4] GORGO is not an acronym. The name GORGO was taken from the name of a 1961 British science-fiction monster movie patterned after the Godzilla and King Kong movies of the 1930s and 1950s.

[5] The current military retirement system is funded by an entry-age normal cost method. The entry-age normal cost percentage, or accrual charge, is the percentage of basic pay that must be contributed over the entire career of a typical group

estimated steady-state accrual cost for "full-time" and for "part-time" personnel, groups reflecting the AC and RC, respectively. The total accrual cost is the sum of these costs. Under the current system, the accrual charge includes the liability associated with the current DB plan, and the DoD Actuary estimated a steady state accrual rate of 44.5 percent in 2016. Under the MCRMC proposal, the accrual charge would include the liabilities associated with the revised DB component of compensation. It would also include the DC plan contributions.

Added to the accrual cost is the continuation pay cost based on the RAND analysis. The overall cost estimate, although not exact, provides a reasonably accurate indication of the proposal's steady-state cost. Cost savings are the difference between the steady-state cost of the proposal and the baseline force. The MCRMC proposal includes a basic continuation pay, paid at YOS 12, with a multiplier of monthly basic pay equal to 2.5 for AC personnel and equal to 0.5 for RC personnel. As mentioned, if the optimized continuation pay multipliers in Table 4.1 exceed these values, the services would need to provide additional continuation pay to sustain the force. From a costing standpoint, we divide the cost of the continuation pay required to sustain retention into the component that is associated with the basic continuation pay and the component associated with the additional continuation pay. This permits us to estimate the additional cost to the services.

Given that data do not permit estimation of the lump sum versus annuity choice and the member's DC contribution choice, and therefore the DoD DC match rate, our approach involves simulating the retention effect of the MCRMC reform under alternative assumptions about the lump-sum choice and the DoD DC match rate. In a footnote, we noted that there are 27 possible combinations of lump-sum and DC plan contribution choices for AC and RC members, for each service, for officers and for enlisted. Rather than simulate steady-state costs under all combinations, we instead report cost estimates on three cases, namely the minimum cost savings case, the maximum cost savings case, and a case that is an example of middle cost savings. The DoD Actuary used these assumptions about the DC plan contribution rate and lump sum versus annuity choice under each case when it produced its steady-state accrual cost estimates for the MCRMC proposal.

In the minimum cost savings case, all members receive an annuity and have a 5-percent DoD DC match rate. In contrast, the maximum cost savings case is where all members receive a 0-percent DoD DC match rate (although DoD still provides the 1-percent automatic contribution) and all enlisted personnel choose a full lump sum. Because of the parameters that have been set for the lump-sum formula, few officers will find the lump sum advantageous relative to the annuity, so for officers, we specify the maximum cost savings case to be where all officers choose the full annuity option (and no lump sum) and have a 0-percent DoD DC match.

The bounds represented by the minimum and maximum cases make extreme assumptions about the DC plan contribution rate and lump-sum choices. For that reason, we provide a middle case example, but other examples are of course possible. Our example of a middle case is where officers and enlisted personnel receive a 3-percent DC match rate and we assume 50 percent of enlisted members choose the annuity, 25 percent choose a partial annuity/partial lump sum, and 25 percent choose a full lump sum with no annuity, while 100 percent of officers choose the annuity.

of new entrants to pay for all future retirement and survivor benefits for that group. The accrual charge is applied to the basic pay bill for the entire force. In contrast to a pay-as-you-go method, this approach means that future retirement costs are incorporated into the computations of the current personnel costs of the force but are not current outlays.

Tables 4.2 and 4.3 present the cost and cost savings estimates for the MCRMC proposal in billions of 2016 dollars. Table 4.2 presents the cost savings to DoD plus the Treasury, while Table 4.3 presents the cost to DoD alone. The Treasury costs include the cost of allowing concurrent receipt of retirement benefits under Combat Related Special Compensation (CRSC) and Concurrent Retirement and Disability Pay (CRDP). CRSC and CRDC are programs that allow eligible military retirees to recover some or all of the retired pay that they usually waive when they receive Department of Veterans Affairs (VA) disability compensation. Before these programs began, receipt of both military retired pay and VA compensation was prohibited, so that those receiving military retired pay faced an offset for the VA disability compensation. These programs allow military retirees to replace some or all of the VA disability offset. CRSC began in 2008 and was created for disability and nondisability DoD retirees with combat-related disabilities. Those who elect to take CRSC must waive VA disability pay, if any. CRDP began in 2004 and restores the retiree's VA disability offset for eligible retirees. The legislation creating CRSC and CRDP assigned their costs to the Treasury, but in effect they are adjustments to the military retirement benefit, and it is therefore appropriate to consider both the DoD retirement accrual charge and the DoD-plus-Treasury accrual charge. For instance, the steady-state annual DoD plus Treasury accrual cost of the baseline force is $24.90 billion for full-time personnel and $1.76 billion for part-time personnel in FY 2016 dollars, according to the DoD Actuary. The total accrual cost is $26.66 billion. CRSC and CRDP benefits depend on the military retirement pay formula. When the formula changes, as would be the case under the MCRMC proposal where the DB multiplier is reduced from 2.5 percent to 2.0 percent, CRSC and CRDP benefits change, so Treasury costs will change as well.

The steady-state accrual cost of the MCRMC proposal under the minimum cost savings case is $21.83 billion for full-time personnel and $1.68 billion for part-time personnel (Table 4.3). These are, respectively, $3.07 billion and $0.08 billion less than baseline costs. In the minimum cost savings case, basic continuation pay costs are $0.30 billion for full-time personnel and $0.03 billion for part-time personnel, while additional continuation pay across the force at YOS 12 to sustain retention is $0.45 billion for full-time personnel and $0.12 billion for part-time personnel. Totaling the numbers, total steady-state accrual cost in the minimum cost savings case is $23.51 billion, or $3.15 billion per year less than the baseline accrual cost. After allowing for both basic and additional continuation pay, the net annual steady-state cost savings are $2.25 billion.

The results for the minimum cost savings case show that even with extreme assumptions, the effects on cost of decreasing the DB annuity and moving compensation forward in the form of continuation pay are sufficient to produce cost savings while still sustaining AC retention and RC participation, despite the addition of the DC plan, even when we assume that all members receive the maximum DC plan match rate.

Given the finding that there is a cost savings in the minimum cost savings case, it is unsurprising that we find cost savings in the maximum cost savings case of $7.69 billion annually. Again, it is important to recognize that this case has unrealistic assumptions about the lump-sum choice and the DC plan contribution rate.

The middle case is intended as an example of mid-range cost savings. Here, we find that the total steady-state accrual costs would fall from $26.66 billion to $21.31 billion, for cost savings of $5.35 billion. Continuation pay to sustain retention would cost a total of $1.04 bil-

Table 4.2
Steady-State DoD and Treasury Cost Savings (FY 2016 $Billions)

	Baseline	Minimum Cost Savings Case	Maximum Cost Savings Case	Middle Cost Savings Case (Example)
Full-time members (AC)				
Normal cost percentage	44.5%	39.0%	29.3%	35.3%
Steady state	$24.90	$21.83	$16.42	$19.75
Steady-state change	–	–$3.07	–$8.47	–$5.15
Basic continuation pay costs	–	$0.30	$0.30	$0.30
Additional continuation pay costs	–	$0.45	$0.67	$0.54
Total change		–$2.32	–$7.50	–$4.30
Part-time members (RC)				
Normal cost percentage	25.9%	24.7%	20.1%	22.9%
Steady state	$1.76	$1.68	$1.37	$1.55
Steady-state change	–	–$0.08	–$0.39	–$0.20
Basic continuation pay costs	–	$0.03	$0.03	$0.03
Additional continuation pay costs	–	$0.12	$0.17	$0.15
Total change		$0.07	–$0.19	–$0.03
Total (AC + RC)				
Steady state	$26.66	$23.51	$17.79	$21.31
Steady-state change	–	–$3.15	–$8.87	–$5.35
Basic continuation pay costs	–	$0.33	$0.33	$0.33
Additional continuation pay costs	–	$0.57	$0.84	$0.69
Total change		–$2.25	–$7.69	–$4.33

NOTE: The normal cost percentage for the three cases shown in the table includes both the cost of the DB portion of the blended system and the cost of DC plan contributions. Insofar as the DC plan contributions are paid directly by DoD to the Thrift Savings Plan Retirement Board rather than to the military retirement fund, the normal cost percentages would be lower than shown in the table, and the table would need to include a separate line that explicitly accounts for the DC contributions.

lion across the AC and RC force.[6] Thus, the total change in the middle case example would be $4.33 billion annually.

The MCRMC final report reports a steady-state cost savings to DoD of $1.8984 billion in the year 2055 (MCRMC, 2015, p. 256). This figure corresponds to the middle cost savings case example. Furthermore, it is for the AC only and reflects the DoD cost savings and not the additional savings to the Treasury. For comparison's sake, we present DoD cost savings (without the addition of the Treasury savings) in Table 4.3. Here, we report an AC steady-state cost savings of $2.29 billion for this example. The reason for the difference between this estimate and what is reported in the MCRMC final report ($1.8984 billion) is that the estimates in

[6] The sum of continuation pay costs in the table is $1.02 billion, differing from the $1.04 billion stated in the text because of rounding error.

Table 4.3 use the estimated steady-state military retirement fund savings provided by the DoD Actuary and add continuation pay costs as provided by the DRM. In contrast, the MCRMC report used the steady-state cost savings provided by the DRM transition model and shown in Figure 5.6b in the next chapter. The transition model provides somewhat different estimates of cost savings in the steady state than the Actuary because of differences in the underlying retention profile used to compute cost savings. The Actuary uses a single retention profile for all services, for officers and for enlisted personnel, developed from cross-sectional data over many years of year-to-year continuation rates, by years of service. The DRM uses separate retention profiles by service and separately for officers and for enlisted personnel within each service, simulated using model estimates based on tracking entry cohorts of military personnel over 20 years. The aggregate of the DRM profiles differs somewhat from the single profile used by the DoD Actuary, leading to differences in the retirement accrual cost change, and therefore differences between the cost savings reported in Table 4.3 and in the MCRMC final report.

Table 4.3
Steady-State DoD Cost Savings (FY 2016 $Billions)

	Baseline	Minimum Cost Savings Case	Maximum Cost Savings Case	Middle Cost Savings Case (Example)
Full-time members (AC)				
Normal cost percentage	31.4%	28.9%	21.0%	25.8%
Steady state	$17.58	$16.16	$11.75	$14.45
Steady-state change	–	–$1.42	–$5.83	–$3.13
Basic continuation pay costs	–	$0.30	$0.30	$0.30
Additional continuation pay costs	–	$0.45	$0.67	$0.54
Total change		–$0.67	–$4.86	–$2.29
Part-time members (RC)				
Normal cost percentage	22.9%	22.3%	17.9%	20.6%
Steady state	$1.56	$1.52	$1.22	$1.40
Steady-state change	–	–$0.04	–$0.34	–$0.16
Basic continuation pay costs	–	$0.03	$0.03	$0.03
Additional continuation pay costs	–	$0.11	$0.17	$0.15
Total change		$0.07	–$0.14	–$0.02
Total (AC + RC)				
Steady state	$19.14	$17.68	$12.97	$15.85
Steady-state change	–	–$1.46	–$6.17	–$3.29
Basic continuation pay costs	–	$0.33	$0.33	$0.33
Additional continuation pay costs	–	$0.57	$0.84	$0.69
Total change		–$0.57	–$5.00	–$2.27

NOTE: The normal cost percentage for the three cases shown in the table includes both the cost of the DB portion of the blended system and the cost of DC plan contributions. Insofar as the DC plan contributions are paid directly to DoD to the Thrift Savings Plan Retirement Board rather than to the military retirement fund, the normal cost percentages would be lower than shown in the table, and the table would need to include a separate line that explicitly accounts for the DC contributions.

CHAPTER FIVE

Transition Results

Given military careers of up to 30 years—and, in some cases, longer—it would take at least 30 years to reach the new steady state as a result of a policy change. The MCRMC was also interested in the effects of its proposal in the transition to the steady state—i.e., during the 30-year period before the new steady state is reached.

A common implementation strategy is to "grandfather" existing members and retirees, so only new entrants are covered by any policy change. Grandfathering is often desirable because policymakers do not want to break the implicit contract with existing members and want to ensure that "promises are kept." The statute creating the MCRMC stated that currently serving members would be grandfathered under the existing system but could be allowed to "opt in" to the new compensation system. The opt-in approach has two potential advantages: Faith is not broken because those who decide to change do so only if they expect to be better off under the new policy, and cost savings can be realized sooner as more members opt in.

The MCRMC requested that RAND use an extended version of the DRM it developed in past research to assess the reform during the transition phase in terms of the effects on retention, costs, and outlays, assuming currently serving military members were permitted to opt in. The extended DRM is detailed in Asch, Mattock, and Hosek (2013) and provides results on the effects for the AC only.

As mentioned in Chapter Three, we assume currently serving members are given a one-year window to opt in to the new system. We assume members will opt in if the expected value of staying in the AC is greater under the new system than the current system. When we simulate the transitional effects on retention, costs, and outlays, we track the retention behavior of each cohort of currently serving members, as well as new entrants who are automatically placed into the new system, and do so in calendar time. A member's YOS in the year the new system is implemented defines his or her cohort under the new system, and the number of individuals in that cohort who opt in to the new system depends on the features of the new system and personal choice, which the simulation handles. The simulation keeps track of each individual's retention experience under the baseline and going forward under the new system, given the individual's opt-in decision. So, the simulation keeps track of individuals in each cohort as they move through their career as well as calendar time, and it aggregates across individuals in different cohorts to produce the cross-sectional force profile for each calendar year. The cross-sectional profiles, together with information on cost of different elements of the reform, are inputs to our computations of changes in cost and outlays during the transition. Because the simulation tracks cohorts by calendar year, we are also able to compute the percentage of a cohort that opts in by YOS and track the retention of that cohort over time.

The chapter begins with results on the percentage of currently serving members who opt in to the new system and then discusses cost savings over the transition years. We do not model opt in among the current population of military retirees. The results on percentage opting in and on cost savings are for the three cases defined in Chapter Four: the minimum cost savings case, the maximum savings case, and a middle example. The final part of the chapter discusses the results on the change in Treasury outlays during the transition period. We also consider three cases in this discussion: the minimum outlay change case, the maximum outlay change case, and a middle case example. As we discuss there, the definitions of the minimum and maximum outlay change cases differ from the definitions of the minimum and maximum cost savings cases. Thus, it is important to recognize that the cases differ for cost savings versus outlay changes.

Percentage of Members Who Opt in

Figure 5.1 shows the percentage of each cohort that is estimated to opt in during the one-year open window to the MCRMC reform proposal under the minimum, middle, and maximum cost savings cases, where cohort is defined as the member's YOS when the policy was implemented. Thus, "YOS cohort = 10" means the member reached YOS 10 at the time of the policy change. Figure 5.1 shows the percentage of the cohort that opts in for the overall AC force, across service and across officer and enlisted personnel, using the FY 2013 AC end strength figures as weights. There are differences across services and across officer and enlisted personnel in the percentage that opt in, though those results are not shown. The results differ because continuation pay differs, and further, for the opt-in choice, the estimated personal discount

Figure 5.1
Percentage of Personnel Who Opt in Under Alternative Cost Savings Cases, by YOS Cohort

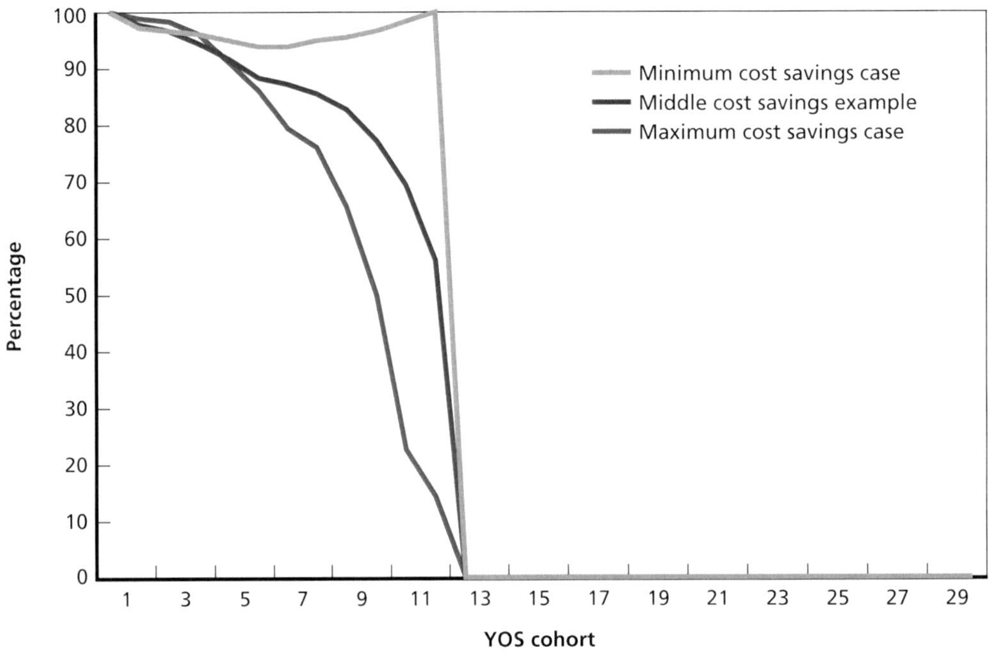

RAND *RR1022-5.1*

rate differs. When the estimated discount rate is higher, as is the case for enlisted personnel, members place a higher value on cases that put more dollars in current pay.

One finding is consistent across cases: The percentage of members who opt in after YOS 12 is zero. The MCRMC reform includes continuation pay that is a lump sum paid out at YOS 12, but that requires pro-rated repayment for members who do not complete an additional four years of service. Members who opt in after YOS 12 will not receive continuation pay, a key element of the reform package. Consequently, no member beyond YOS 12 opts in.[1]

In the minimum cost savings case, the most-junior members are virtually 100-percent likely to opt in. Those in YOS cohort 7 opt in at a rate of 80 percent, while those in cohort 10 opt in at a 50-percent rate. In the maximum cost savings case, virtually 100 percent of all members with 12 or fewer years of service opt in. This case provides the minimum DC match rate of 0 percent and assumes all enlisted members who qualify for AC retirement receive a full lump sum, in place of the second career annuity.

As mentioned in earlier chapters, the minimum and maximum cost savings cases are bookend cases to illustrate the range of potential effects, but they are not realistic in terms of likely behavior. The middle case is an example. It assumes that all members receive the default 3-percent DC match rate, with 50 percent choosing an annuity, 25 percent choosing a partial annuity/partial lump sum, and 25 percent choosing a full lump sum between retirement and age 67 for those in the AC. Here, we find that the percentage of currently serving members that would participate in the new plan is at least 50 percent among those with 12 or fewer years of service. As with the minimum cost savings case, the fraction increases among those with fewer YOS.

The opt-in rates correspond to the continuation pay multiplier across different DC match rates. The multiplier is highest when the match rate is zero, lower for a 3-percent match, and lower still for a 5-percent match (Table 4.1). These rates in turn correspond to the maximum, middle, and minimum cost savings. This correspondence suggests the importance of the continuation pay amount as an inducement to opt in.

Retention During the Transition

We can also compute the change in retention during the transition period, but, as mentioned in Chapter Three, the reform proposal is designed to sustain retention. Consequently, retention is by and large constant in the transition to the steady state, as well as in the steady state. We show this pattern in this subsection.

Because the continuation pay is optimized to sustain retention, it is no surprise that, in conjunction with the other features of the MCRMC proposal, force size and shape remains virtually the same under the reform. This is the case both in the steady state, as shown in Chapter Four, and in the transition to the steady state, as illustrated in Figure 5.2 for Army enlisted personnel. The figure assumes a 3-percent DC match rate and all qualified members choose an annuity between retirement and age 67. The figure appears to show only one line—the retention profile for Army enlisted personnel—but in fact it shows 30 separate lines. Each line represents the retention profile in year s, where s is the number of years since the policy

[1] We mentioned earlier that we did not model opt in among current military retirees. However, the fact that no currently serving members opt in after 12 years of service suggests that few, if any, retirees would opt in.

Figure 5.2
Simulated Army AC Enlisted Retention in Transition to Steady State, 3-Percent DC Match Rate and Annuity-Only Choice

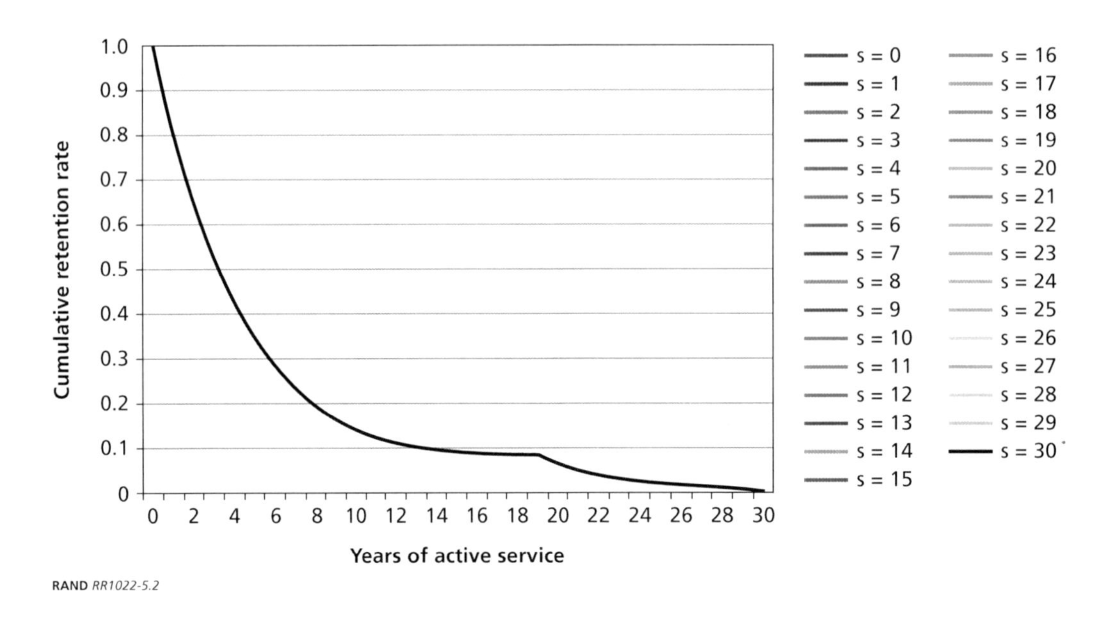

RAND *RR1022-5.2*

change occurred. Year 0 is the steady state in the baseline, and year 30 is the new steady state. The fact that all of the lines are identical in Figure 5.2 means that retention is unchanged in the transition period.

The composition of personnel changes as time elapses in terms of who is and is not covered by the new policy, as shown in Figures 5.3 and 5.4. In the baseline ($s = 0$), all members are covered by the current compensation system and none are covered by the new system, hence the flat blue line along the axis in Figure 5.3, which is for members who opt in, and the baseline retention curve in Figure 5.4, which is for members under the current system at $s = 0$. In the first year ($s = 1$), new entrants are covered by the new system, while everyone else is under the current system and has the option to remain in the current system or to switch to the new system. Figure 5.3 shows the retention profiles over calendar year (or year since reform) of only those in the new system. Consistent with the results in Figure 5.1, only members with 12 or fewer years of service choose to opt in to the new system, so at s = 1, we observe retention only through YOS 12 for those in the new system. Figure 5.4 shows the retention profiles by YOS and years since reform for those who remain under the current system. For the most part, only those with more than 12 years of service stay in the current system and, as time elapses, these individuals flow out of the military. In the steady state, no members are under the current system, and all members are under the new system. Thus, while the retention profile in Figure 5.2 remains unchanged in the transition phase, the composition of members in terms of those under the new system versus the current system changes as time elapses.

Cost Savings in the Transition

The analysis of costs focuses on current cash compensation and retirement/deferred compensation costs. Because the force size and shape remain virtually unchanged, changes in current

Figure 5.3
Simulated Army AC Enlisted Retention of Members Who Opt in, Transition to Steady State, 3-Percent DC Match Rate and Annuity-Only Choice

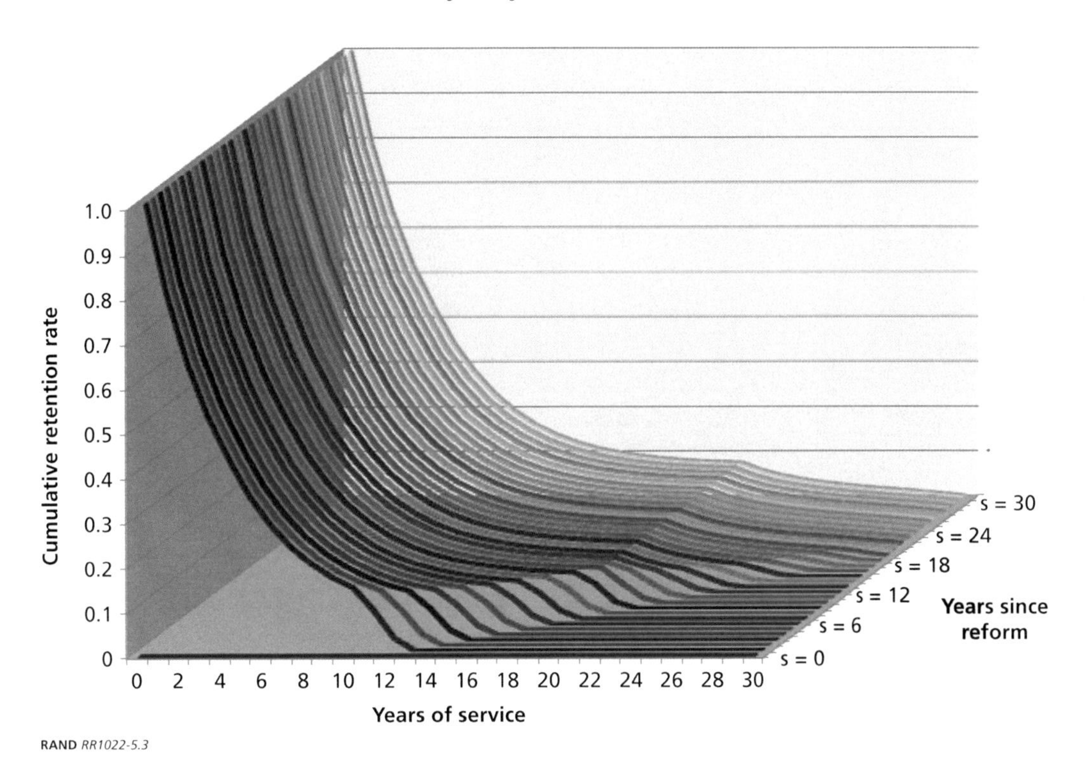

RAND RR1022-5.3

Figure 5.4
Simulated Army AC Enlisted Retention of Members Who Do Not Opt in, Transition to Steady State, 3-Percent DC Match Rate and Annuity-Only Choice

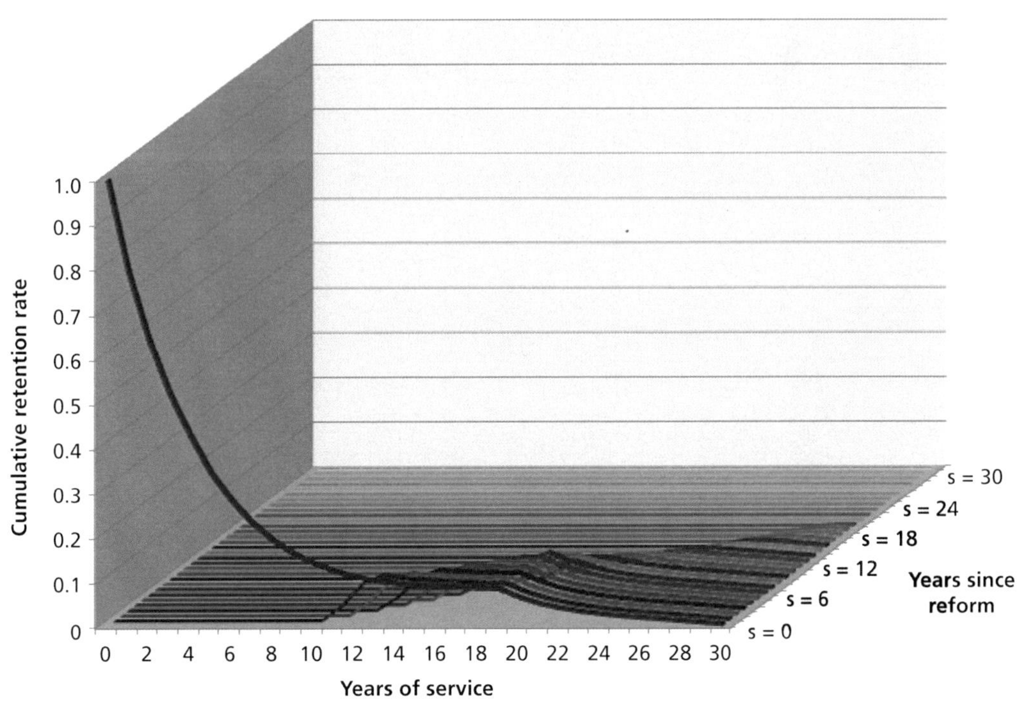

RAND RR1022-5.4

compensation costs include only the new element of cash compensation, namely continuation pay. The cost of basic pay and the other elements of regular military compensation do not change when the force size and experience mix is the same. Changes in deferred compensation costs reflect changes in the retirement accrual charge that result under the MCRMC proposal.[2]

As discussed in Chapter Four, the current military retirement system is funded by an entry-age normal cost method. The entry-age normal cost percentage is the percentage of basic pay that must be contributed over the entire career of a typical group of new entrants to pay for all future retirement and survivor benefits for that group. The normal cost percentage is then applied to the basic pay bill for the entire force. In contrast to a pay-as-you-go method, this approach means that future retirement costs are incorporated into the computations of the current personnel costs of the force. The accrual cost is not an outlay but a nominal payment into the Military Retirement Fund, which is managed by the Treasury. Under the current system, the accrual charge includes the liability associated with the current DB plan and is estimated to be 44.5 percent in 2016 by the DoD Actuary. The accrual charge will include the liabilities associated with the revised DB component of compensation and, for the purpose of our analysis, the DC plan. Also, although the lump-sum payments increase current compensation, they would be funded through the normal cost percentage.

The new accrual charge would decrease under the MCRMC proposal. This is because deferred retirement benefits decrease while current compensation in the form of continuation pay increases. Current compensation is further increased if members choose a lump sum over an annuity between retirement and age 67. These changes result in a net decrease in the cost of retirement-related compensation, which in turn implies that a smaller percentage of basic pay needs to be set aside to fund the benefits—thus the normal cost percentage also decreases. The amount of the cost savings depends on how much the accrual charge decreases and how much pay is moved forward into current compensation. This chapter has shown the steady-state cost savings under three cases, two of which bookend the possible range of cost savings and one that is an example of a middle cost savings case. Here, we show the path of cost savings in the transition period for these three cases.

The green line in Figure 5.5 shows the time patterns for the AC of the change in personnel costs in FY 2016 dollars to DoD and the Treasury under the minimum cost-savings case. The red and blue lines show the corresponding time patterns of cost savings for the middle example and for the maximum cost savings case, respectively. In this figure, personnel are allowed to opt in to the new system.

In each case, personnel costs decrease relative to the baseline in the first year that the new policy is implemented. In the minimum cost savings case, cost savings one year after the policy change are $1.31 billion. This represents the $1.53 billion drop in the Military Retirement Fund accrual charge for both DoD and the Treasury, but a $250 million increase in costs from continuation pay. In the maximum cost savings case, cost savings in the first year are $3.98

[2] For the purpose of our analysis and for this report, the costs of the DC plan contributions were included in the retirement accrual charge, so the cost of those contributions is included in the decrease in the accrual charge associated with the lower DB multiplier. That is, we implicitly assumed a "pass-through," whereby DoD paid out the DC contributions to the military retirement fund as part of its accrual charge and Treasury paid those contributions to the Thrift Savings Plan Retirement Savings Board. Alternatively, the DC plan contributions might be paid directly by DoD to the Thrift Savings Plan Board, without going through the military retirement fund. If this were the case, the new retirement accrual charge would include only the DB portion of the blended system and would differ from what is shown in Tables 4.2 and 4.3. Furthermore, the cost decrease associated with normal cost percentage shown in the tables would differ as well.

Figure 5.5
Change in DoD and Treasury Personnel Costs for AC Personnel Under Three Example Cases (FY 2016 $Millions)

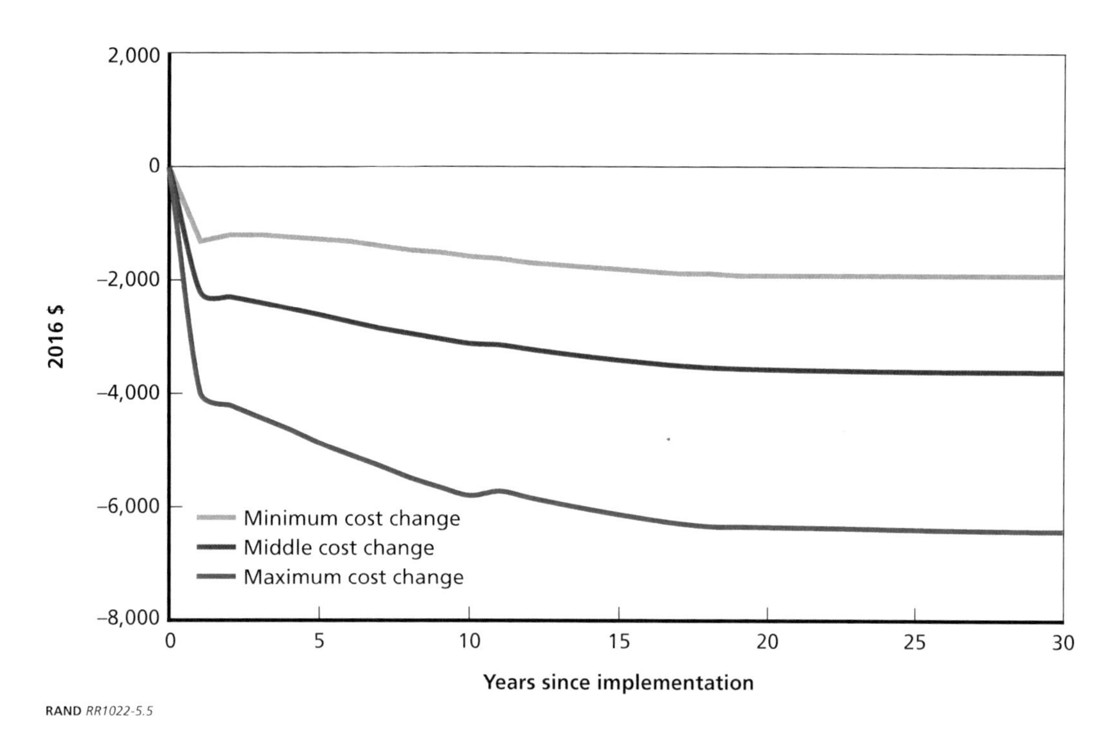

billion, representing a $4.73 billion cost savings associated with the accrual charge change but an $840 million increase in costs from continuation pay. In the middle example, cost savings in the first year are $2.21 billion.

Cost savings are larger in the first year, when currently serving members are permitted to opt in to the new system, than in the case when they are not given the opt-in choice. As shown in Figure 5.1, nearly 100 percent of currently serving members with 12 or fewer years of service opt in under the maximum cost savings case, while a lower percentage opt in under the minimum cost savings case. Nonetheless, in both cases, over one-half of the steady-state cost savings are realized in the first year when opt in is permitted.

In previous analysis (Asch, Hosek, and Mattock, 2014) we compared the time pattern of cost savings when members are not permitted to opt in with the pattern when they are permitted. We found that costs decrease slightly in the first year in the absence of the opt-in choice because both DoD and the Treasury realize a cost savings associated with the new entrants under the new, less costly system. The decrease occurs slowly as new entrants are put under the proposed, less costly compensation system and existing members under the current system gradually age and eventually leave service. Costs continue their downward path until the new steady state is reached. Cost savings are realized earlier when currently serving members can opt in, and the higher the percentage of currently serving members that opt in, the faster the steady-state cost savings are reached.

Figure 5.6 shows the corresponding cost savings figures to DoD only. We present these results to show how our results compare with the DoD cost savings reported in the MCRMC final report. As mentioned in Chapter Four, the MCRMC reported steady-state savings to

Figure 5.6
Change in DoD Personnel Costs for AC Personnel Under Three Example Cases (FY 2016 $Millions)

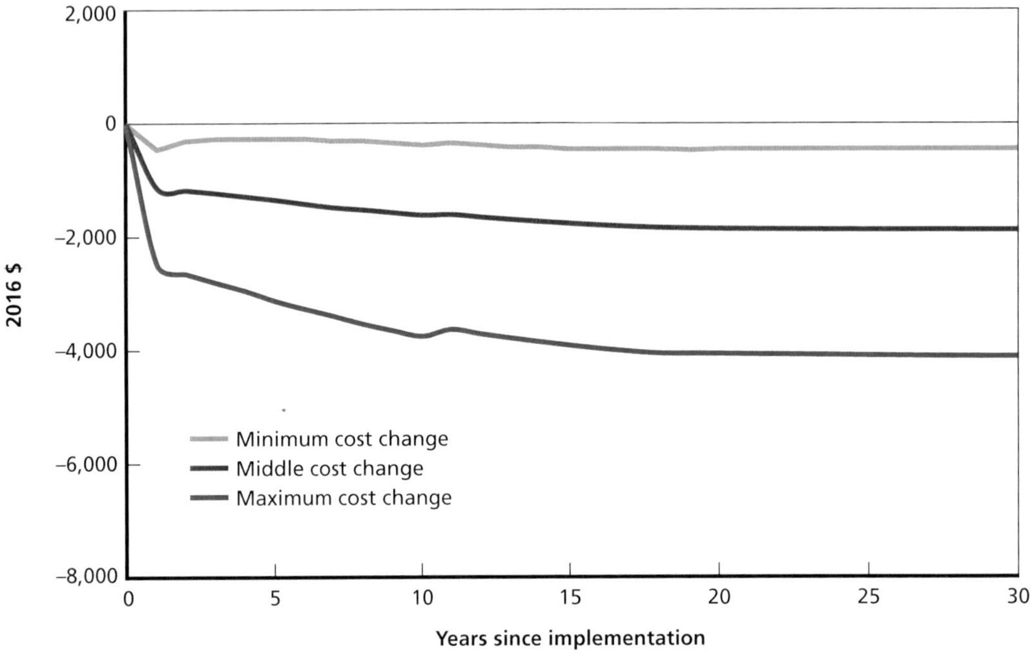

RAND *RR1022-5.6*

DoD of $1.894 billion. This amount corresponds to the steady-state cost savings shown in the red line of Figure 5.6, after all members have spent an entire career under the new system.

Changes in Treasury Outlays in the Transition

Outlays are current-year government expenditures. Outlays increase under the MCRMC proposal when the government provides continuation pay, contributes to the DC plan on behalf of members, and provides lump-sum payments to separating members. However, outlays for the DB plan decrease under the proposal. The time pattern of the change in costs and outlays in the transition period will depend on the DC plan contribution rates chosen by members, their lump-sum choices, the level of continuation pay, and the percentage of members who opt in to the new system.

We consider three cases of outlay changes, namely the minimum outlay change case, the maximum outlay change case, and a middle outlay change case. The specifics of the minimum and maximum outlay change cases *differ* from the minimum and maximum cost savings cases considered earlier. The reason is that the minimum cost savings case is not also the minimum outlay change case, and similarly for the maximum cost savings case. The minimum outlay change case is where we assume all members receive a 0-percent DC match rate and all members choose an annuity rather than a lump sum. In contrast, the minimum cost savings case is the case where we assume all members receive a 5-percent DC match rate and all choose an annuity. This latter case is not a minimum outlay case because the 5-percent DC match results in larger outlays than the 0-percent rate. The maximum outlay change case is the case where all members receive a 5-percent DC match rate and all enlisted members choose a full lump sum over an annuity. As explained, we assume all officers choose an annuity. The middle outlay change case is the same as the middle cost savings case; we assume all members receive a 3-percent DC match rate, 50 percent of enlisted personnel choose the annuity, 25 percent choose a partial annuity/partial lump sum, and 25 percent choose a full lump sum, while all officers choose an annuity. The bounds represented by the minimum and maximum cases make extreme assumptions about the DC plan contribution rate and lump-sum choices, so the middle case might be more representative of actual behavior.

Figures 5.7 and 5.8 show the time pattern for the change in Treasury outlays on behalf of AC personnel when existing members can opt in and new members are automatically placed under the MCRMC reform, for the minimum, middle, and maximum outlay change cases. To help explain the patterns observed in these figures, Figure 5.9 shows the time pattern of outlays by element of outlay—DC plan contributions, continuation pay, lump sum, and the DB plan, for the middle outlay change example.

The general pattern of the change in outlays is that, initially, Treasury outlays increase above the baseline in the years after the policy is implemented and then eventually decrease below the baseline. Figure 5.10 helps explain why this is the case. In the first seven years, outlays increase in the middle example because of DoD's DC plan contributions for new entrants and currently serving members who opt in to the new system (the red bars).[3] In addition, outlays occur because of continuation pay payments to opted-in members who reach YOS 12

Figure 5.7
Change in Treasury Outlays for AC Personnel (FY 2016 $Millions)

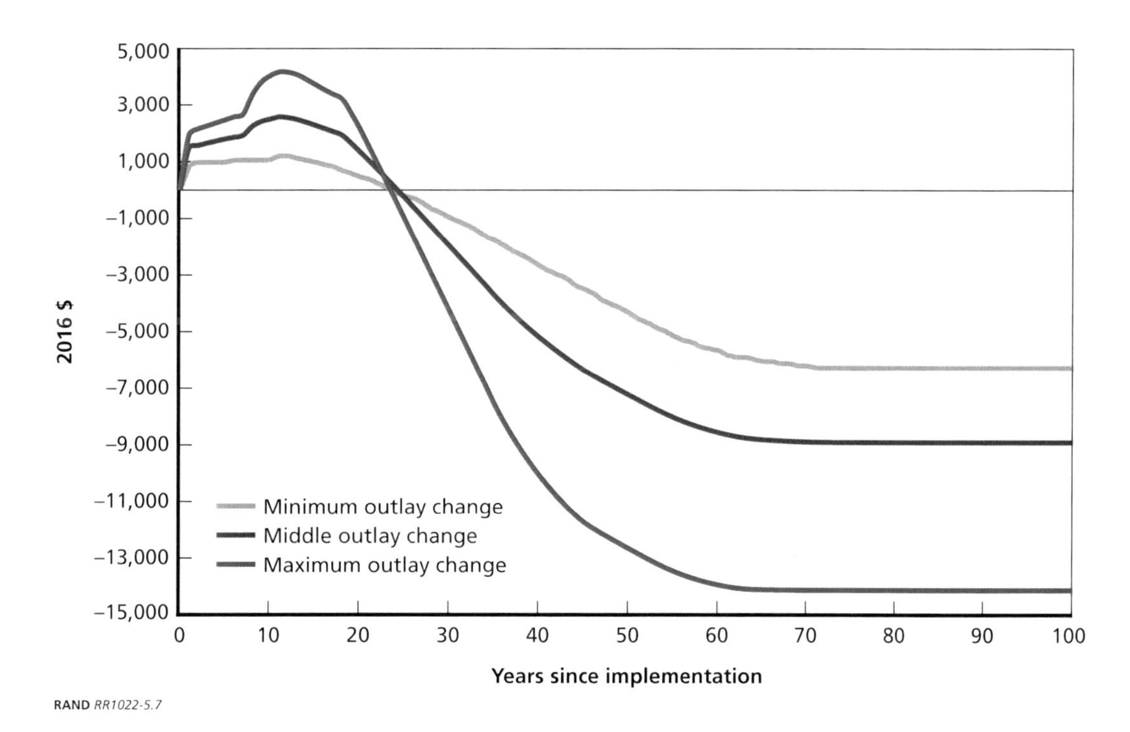

RAND *RR1022-5.7*

[3] DC plan contributions occur between YOS 3 and YOS 20. Although nonvested members have no claim on these contributions, they revert back to the Treasury (cf. Chapter 84 of Title 5, United States Code, sections 8432 and 8437, pertaining to forfeit contributions to the Thrift Savings Fund). Thus, Treasury outlays for DC plan contributions in our analysis begin when members become vested or after YOS 3.

Figure 5.8
Change in Treasury Outlays for AC Personnel, by Element of Outlay for Middle Outlay Change Case (FY 2016 $Millions)

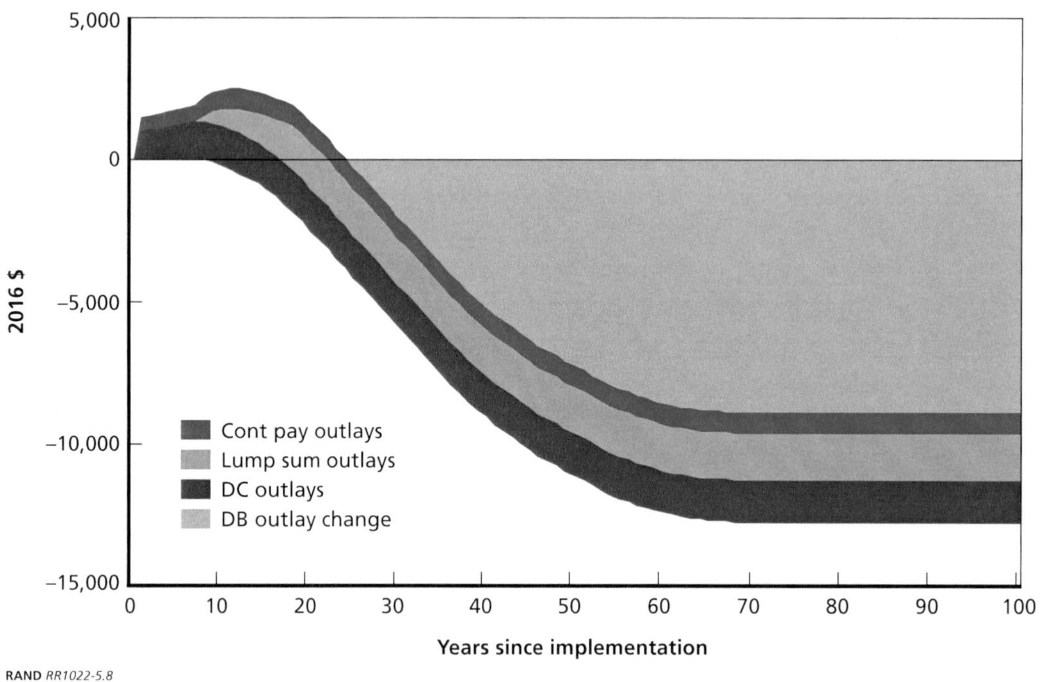

(the blue bars). However, because only those with 12 or fewer years of service opt in, no DB or lump-sum outlays are made until eight years after the policy change, when members who had reached YOS 12 at the time of the policy change first become eligible for the DB. Given our assumption in the middle case that one-half of enlisted members receive either a full or partial lump sum, Treasury outlays as a result of the lump-sum payments (the green bars) begin in year 8. The DB outlay change is reflected by the gray area. The change in DB outlay is always a negative number because outlays for the new DB plan are lower under the MCRMC proposal than for the current system. The other three elements of cost, which are in red, green, and blue, are positive and stack on top of the DB outlay. Thus, the full decrease in the DB outlay is at the bottom of the red bar; then the other three elements stack on top of the decreased DB outlay, and the top of the blue bar indicates the net decrease in outlays.

By design, the minimum outlay change case has the smallest initial increments in Treasury outlays and the smallest ultimate decrease in outlays. The maximum case has the largest initial increment, and the largest ultimate decrement in outlays. Specifically, in the first year of the policy change, we find that outlays would increase by $900 million in the minimum case, rising to $1.2 billion after 11 years and then decreasing, becoming negative in year 24 (–$60 million) and ultimately decreasing to –$6.30 billion. The new steady state occurs after 75 years, when all members under the current system have died and the Treasury is no longer paying annuities based on the current system.

In the maximum outlay change case, outlays in the first year would increase by $1.92 billion, growing to $4.16 billion in year 11, then decreasing in year 24 (–$420 million) and reaching a steady-state drop of –$14.14 billion. In the middle example, initial outlays would increase by $1.47 billion, up to $2.57 billion in year 11, become negative in year 24 (–$290 million), and reach a steady-state drop of –$8.90 billion. The steady state occurs in the middle case after 75 years, corresponding to the year 2090. The MCRMC final report shows the changes in federal outlays through 2055 (MCRMC, 2015, p. 256), with a decline of $4.666 billion in that year, corresponding to about one-half of the ultimate $8.9 billion drop in outlays in the steady state.

Transition Period for the Reserve Component

In this section, we present a qualitative analysis of the likely effects for the RC during the transition period, drawing on what we learned from the DRM analysis for the AC. We have not extended the transition modeling in the DRM to the reserves to produce estimates of the retention, cost, and outlay effects for the RC. Our observations for the RC are based on the steady-state results for the RC and the patterns, direction, and magnitude of the transition effects for the AC. Thus, our conclusions about the RC are informed observations.

Our steady-state results showed that RC participation is close to baseline participation regardless of the DoD matching rate and regardless of the lump-sum option choice. Given the parameters of the DC plan and DB annuity (multipliers, vesting, age of benefit receipt) and the parameters of continuation pay and lump sums (formula, when paid), the optimized multipliers for continuation pay and the lump sums perform well to sustain RC participation. With these results in mind, we can consider what might be the likely effects for the RC in the transition phase.

RC participation is quite close to baseline, especially for the annuity case. Thus, like the AC, we expect RC participation in the transition period to be much the same as at baseline. In the steady state, when individuals choose one of the lump-sum options we find that there are some deviations relative to the baseline. Specifically, there is a small decrease in pre–YOS 20 participation, a decrease in post–YOS 20 participation, and continuation pay does not restore the experience mix. However, as noted in a footnote, this deviation could be an artifact of the assumption that all members make the same lump-sum choice and have the same DC plan contribution rate, or the fixed parameters defining the lump-sum values, or both. Insofar as they result from the fixed parameters, there could be some years in the transition period when RC strength is lower than the baseline. The timing of these effects will depend on the propensity of RC members to opt in to the new system.

The pattern of RC cost change will depend on the extent to which individuals choose to take one of the lump-sum options and on the contribution rate for the DC plan. Table 4.2 shows that the cost change varies from a $70 million cost increase to a $190 million cost decrease relative to baseline. If there are cost savings, we would expect to see a pattern similar to that observed for the AC. For all of the alternatives, during the transition phase the cost savings for the AC would dominate any cost change for the RC. Similarly, like outlays for the AC, outlays on behalf of RC members will likely increase initially during the transition period because of DC plan contributions and continuation pay costs.

CHAPTER SIX

Concluding Thoughts

Overview of Findings

The MCRMC proposal can support a steady-state force and experience mix closely equivalent to the current force. While there was no presumption that future requirements will call for the same size and mix as the baseline force, a criterion for assessment was that any reform alternative could achieve the baseline. This finding is of fundamental importance to assure the viability of the proposal: It can feasibly recreate force size and nearly create force shape. The proposal brings compensation forward by decreasing deferred compensation in the form of defined benefits and increasing current compensation in the form of continuation pay, and if members so choose, lump-sum payments. The proposal also provides a DC plan vesting at the start of YOS 3, assuring that a much higher percentage of service members will have a retirement benefit as compared with the current system. It also provides an incentive to continue to serve tantamount to that of the current system.

Moving some amount of compensation forward is beneficial to members by providing them with more resources earlier in their career. Keeping a significant, though decreased, portion of compensation in deferred compensation ensures the availability of income between retirement and age 67 as well as throughout old age, in recognition of long years of service and sacrifice.

It is also beneficial by providing cost savings while maintaining force size and shape. We estimated the minimum and maximum possible cost savings and change in Treasury outlays associated with the proposal, as well as for a mid-range example. Cost savings ranged from $2.3 billion to $7.7 billion per year in the steady state, with a mid-range example saving $4.3 billion per year including the cost savings to both DoD and the Treasury and including the cost savings for the AC as well as the RC. Thus, the lower cost associated with the lower accrual charge more than offsets the higher costs associated with the additional elements, yet still sustains retention relative to the baseline.

These cost-savings figures do not incorporate other benefits of reform. These include the increased value to the member from having a higher likelihood of vesting and from being able to choose the form of the second-career benefit—i.e., full annuity, full lump sum, or partial annuity and partial lump sum—and the increased value to the services from the possibility of more flexibility in managing the force. Also, the modeling and cost savings do not account for other changes that may be recommended by the MCRMC in the areas of quality of life and health care.

Policymakers care about the transition period as well as the steady-state effects of retirement reform in part because of the budget cycle's focus on near-term costs and outlays and, in part, because different implementation strategies can affect the time pattern of the effects of reform. The statute that created the MCRMC opened the door for allowing currently serving members and retirees to opt in to any adopted reform. RAND extended the DRM to consider the extent to which currently serving members would opt in to the new system as well as the retention, cost, and outlay effects for the AC during the transition period.

We find that no currently serving member with more than 12 years of service would opt in to the MCRMC proposal. This is because the proposal provides continuation pay at YOS 12 and those who opt in after that point would miss receiving this payment. We find that participation rates would be highest among the most-junior members, currently serving members, and at least 50 percent among those with 10 or fewer years of service in the three cost savings cases (minimum, maximum, and middle example) we considered.

Because the continuation pays are optimized to sustain retention, the MCRMC proposal sustains retention not only in the steady state but also in the transition phase. As for the time pattern of cost savings, we find in the three cases we consider that over half of the steady-state cost savings are realized in the first year when opt in is permitted. This is because a significant percentage of junior members—who comprise the bulk of the force—would opt in to the new system in those cases we considered, though there is considerable variation in opt-in rates across services and between enlisted personnel and officers.

We find that Treasury outlays initially increase under each alternative because of the additional elements of compensation, specifically the DC plan contributions, the continuation pay, and the possibility of a lump-sum payment in lieu of all or part of the annuity between retirement and age 67. DC plan contributions represent a transfer from the Treasury to the DC fund for each member. Outlays eventually decrease because of the less-generous DB annuity under each alternative. This time pattern differs from the time pattern for cost, where we find that costs decrease from the beginning. Furthermore, the assumptions for the minimum and maximum cost savings cases differ from those for the minimum and maximum outlay change cases. For example, cost savings will be minimized when all members choose an annuity and receive a 5-percent DC match rate. But outlay changes will be minimized when all members choose an annuity and receive a 0-percent DC match rate. The 0-percent match rate decreases outlays but increases cost savings, because the accrual charge is lower when the contribution rate is lower. In the middle outlay change example (which is also the middle cost savings case), initial outlays would increase by $1.9 billion in the first year (in FY 2016 dollars) but eventually fall by $8.9 billion in the steady state relative to baseline, where the steady state would be reached in the year 2090.

In short, we find a trade-off between cost and outlays in the near term. Costs decrease in the initial years, and the drop in costs occurs faster the more current members opt in. Outlays increase in the initial years, and the increase is larger and occurs more rapidly the more current members opt in. In the long run, both costs and outlays are lower than at baseline. Thus, policymakers must weigh the advantages of initial cost decreases against higher initial outlay increases.

Findings Within the Broader Context of Reform

To place our findings in context, we return to the issues that must be balanced when considering compensation reform. Past commissions and reviews critiqued the military compensation system and found it inequitable, inefficient, and inflexible. A Congressional Research Service report (Henning, 2011) offers a succinct statement: The military retirement system is "inefficient because it defers too much compensation until the completion of a military career; inflexible because it does not facilitate force management or encourage longer careers; and inequitable because most service members never qualify or vest." Yet the system has functioned to help meet manning requirements and, specifically, it has helped stabilize the retention of mid-career personnel who bring considerable training, experience, and leadership and who ultimately make up the pool of candidates for top leadership positions—and whose abrupt departure would pose considerable downside risk to capability and cost. It also has advantages from the perspective of the service member. It provides a low-risk and predictable source of income for old age and, by being available immediately upon separation, the retirement annuity can assist in the transition to a second career.

A reform must conserve the strengths of the current system and yet address its deficiencies. Not surprisingly, this implies a class of reforms that are a blend of DB and DC systems and with less deferred compensation and more current compensation. A DB plan provides a predictable source of income and a transition benefit, while a DC plan has the advantage of being portable and provides individual choice and flexibility regarding how plan funds are invested. Earlier vesting of either or both of the DB and DC plans can improve equity by increasing the likelihood that a given entrant will become vested. A blended plan can also accommodate the services' needs for an efficient and flexible force management tool. Depending on how retired pay is computed in the DB element, how retirement eligibility criteria are defined, and when payouts are made, the system can be designed to induce members to stay until certain career points and then induce them to leave when desired. As mentioned, less deferred compensation and more current compensation can reduce costs while sustaining retention, and targeting current compensation can help shape career retention profiles across communities.

The MCRMC proposal is a blended plan that keeps a substantial portion of compensation as deferred compensation in a DB plan and keeps vesting for this plan at YOS 20, but a portion of deferred compensation is brought forward in the form of continuation pay and the lump-sum options. Its DC plan vests much earlier, at YOS 3, and pays out in old age. While we do not analyze the behavioral aspects of matching contributions, past studies show that automatic enrollment in a DC plan with a default contribution rate can dramatically increase participation in the retirement plan, particularly among those who were least likely to participate in a retirement saving plan without these features (Madrian and Shea, 2001). Our results also show that the MCRMC proposal can reproduce the baseline AC and RC force size. It also produces a gain in efficiency as a consequence of bringing some compensation forward by obtaining a force of the current size and generally the same shape at lower cost.

The MCRMC proposal includes features to promote more flexible management of personnel and to vary career lengths across communities of personnel. Our analysis implied that different values of continuation pay multipliers could alter the size and experience mix of the force if desired. In our simulations, continuation pay multipliers differed between services and between officer and enlisted communities within a service, and, at the request of the MCRMC, continuation pay included fixed basic continuation pay multipliers of 2.5 for AC personnel and 0.5 for RC personnel. Still, given legislative authority, a service could offer continuation pay at different rates by occupation and thus shape the occupation retention profiles of its officer and enlisted communities. The MCRMC recommended that the Secretary of Defense be given authority to change the service requirement to qualify for DB plan benefits to facilitate the shaping of the force to achieve longer or shorter careers within occupational specialties or within specific communities.

Findings Within the Context of the DoD Concepts for Modernizing the Retirement System

The 2014 DoD white paper discussed two concepts for modernizing the military retirement system (U.S. Department of Defense, 2014). As required by law, these concepts were transmitted to the MCRMC for their deliberation. RAND analyzed the effects on AC retention, RC participation, cost, and outlay of these concepts (Asch, Hosek, and Mattock, 2014), and both the RAND study and the white paper provide details of the concepts. In brief, concepts would replace the current system with a system that includes three elements, though the details of these elements differ between Concepts I and II.

1. Modify the current DB system to reduce the amount of payout to the member, though both concepts would maintain vesting at YOS 20, with an immediate payout. The two concepts differ in how they would change the amount and the timing of the payout.
2. Supplement the DB plan with a DC plan that would vest earlier than YOS 20, specifically vest after YOS 6; would involve automatic contributions by DoD on behalf of the member equal to 5 percent of his or her basic pay; and would begin the payout of the benefit as early as age 59½.
3. Increase current cash compensation, in the form of supplemental pays, though the amount of these changes in pay would differ between the two concepts. The purpose of these pays is to sustain the size and experience mix of the force.

Under both concepts, the DB plan benefit would still be based on the average of the three highest years of basic pay, a multiplier, and years of service. Both concepts pay a lower retirement annuity than under the current system and compensate for this reduction with supplemental pay that would be in addition to existing special and incentive pays—effectively shifting a portion of deferred compensation to current pay, in addition to the DC plan. Concept I is crafted around a two-tiered DB benefit for both the active and reserve components. The ability to start a second career was the driver for developing a two-tier benefit. The first tier provides a partial retirement benefit during the member's normal second-career years to *both* active and reserve component members. The second tier begins when members are in their early 60s and pays full retirement benefits. Concept I provides RC members with a partial benefit during a second career, whereas under the current system RC members generally are eligible to receive benefits only upon reaching age 60. Concept II offers a single tier of DB benefits for both the active and reserve components like the current retirement system, except the multiplier would be less than its current 2.5 percent. As under the current system, eligible RC members would not generally begin receiving benefits until age 60. Under this concept, the retirement benefit is a full benefit in all years, not a partial benefit in the years before retirees reach their early 60s.

Both concepts would offer two types of supplemental pays—transition pay and continuation pay—though the amount of the pays would differ under the two concepts. The purpose of these pays is to sustain the size and experience mix of the force, given the decrease in the defined benefit. The transition pay would be a multiple of final annual basic pay and would be offered upon retirement to AC members with at least 20 years of service. To protect the funding for transition pay, the multiplier is set to be the same across services and across enlisted and officer personnel. This feature might inhibit the role of transition pay in helping the services reshape the force size and experience mix, if that is desired. Funding for the transition pay is rolled into the retirement accrual charge. The continuation pay would be a multiple of monthly basic pay and targeted to specific years of service to sustain retention. Continuation pay would vary by service, by whether personnel are officers or enlisted, between active and reserve components, and possibly by occupational area. In the analysis conducted during this review, the pay was targeted to enlisted personnel at YOS 12 and officers at YOS 16.

The MCRMC proposal has similarities with the two concepts. It also is a blended plan that includes the same three components, a less-generous DB plan, the addition of a DC plan that requires DoD contributions, and supplemental pay that increases current compensation. Like the two concepts, the MCRMC DB plan retains 20-year vesting in an immediate annuity, and the DC plan vests earlier than 20 years. The MCRMC plan also includes continuation pay to sustain the retention and experience mix of the force, like the two concepts. The MCRMC plan does not include transition pay, per se, but it does include an option for members to choose a partial or full lump sum, with a reduced or zero annuity between retirement and age 67. The MCRMC proposal blends ideas in Concepts I and II. It is similar to Concept II because it has a single tier of DB benefits. Yet, like Concept I, it allows members to receive a partial annuity during the working age (or second career) stage of retirement. It is also like Concept II in terms of the reform features for the RC because it does not include an immediate annuity for RC members but provides RC members with DoD DC plan contributions, supplemental pay in the form of continuation pay, and a less-generous DB plan.

Details of the MCRMC proposal differ from DoD's two concepts. The two concepts do not include the lump-sum choice for either the AC or RC, nor do they stipulate a basic continuation pay amount based on a 2.5 multiplier for AC personnel and 0.5 for RC personnel. The MCRMC proposal would vest the DC plan at YOS 3 rather than YOS 6 and includes a matching feature as well as the opportunity for members to opt-out of the match and receive only a 1-percent automatic contribution.

Conclusion

The 2014 DoD white paper argued that the objectives of efforts to modernize the retirement system should be to provide members who faithfully serve their country with a robust retirement, to provide force managers with tools to maintain and shape the force structure, and to provide the American taxpayers an effective force at an affordable cost. Two other considerations discussed were the need to protect those already retired and those currently serving and to not harm the existing force structure and capability of the force. The paper argued that the two concepts employ a blended plan, and this type of plan offers an approach for developing a modernized military retirement system without the drawbacks of other proposed alternatives (U.S. Department of Defense, 2014). Importantly, the MCRMC proposal falls squarely within the framework and approach offered by DoD.

The findings of our analysis of the MCRMC proposal offer assurance to the military services that current manpower requirements can be met under the MCRMC retirement reform proposal. Such assurance is of fundamental importance for national defense. The MCRMC proposal can also attain force sizes and shapes different from those existing today. In our analyses, we treated continuation pay as force-wide elements of compensation, but in principle a service could differentiate continuation pay to vary career lengths by field or occupation. DoD and the services are also interested in cost savings, and the proposal provides them. Finally, the proposal keeps faith with service members by maintaining a significant, secure retirement benefit, but one restructured to include a defined contribution with early vesting. The proposal requires an initial increase in outlays, although outlays eventually decrease. The increase in the percentage of service members who vest would bring the military retirement system into closer alignment with DC plans offered by private-sector employers. The lump-sum option can provide funds to assist service members in various ways—for example, in the transition to new careers, relocation, making down payments, paying for their children's education, and so forth. The proposal still reserves a large portion of funds for second-career and old-age retirement benefits. Also, the DC funds—following the Thrift Savings Plan model—would be placed in government-approved funds that protect against high-risk or fraudulent investment vehicles. Therefore, the proposal offered by the MCRMC has a number of merits from the perspectives of the military, the government, the taxpayer, and the service member. Some of the merits are generic to blended plans, as seen in the findings of past studies and commissions, while others, such as lump-sum choices, are specific to the MCRMC plan.

APPENDIX A

Steady-State Retention

This appendix shows the steady-state AC retention and RC participation results for Army enlisted personnel when the DoD DC match rate is 0 percent (Figure A.1) and 5 percent (Figure A.2) and when all of the multipliers and multiplier parameters are optimized to restore retention under the reform alternative to the baseline. Figure 4.1 in Chapter Four shows the Army enlisted results for the 3-percent DoD match rate case. In addition, this Appendix shows the AC and RC results for the Navy (Figures A.3–A.5), the Marine Corps (Figures A.6–A.8), and the Air Force (Figures A.9–A.11).

Army

Figure A.1a
Army AC Enlisted Retention and Prior-Active RC Enlisted Participation, 0-Percent DoD Match Rate, Annuity-Only Choice

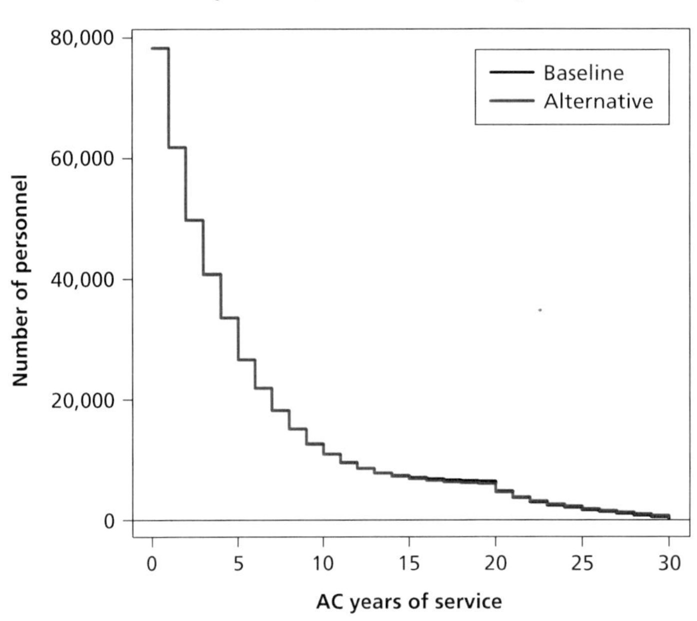

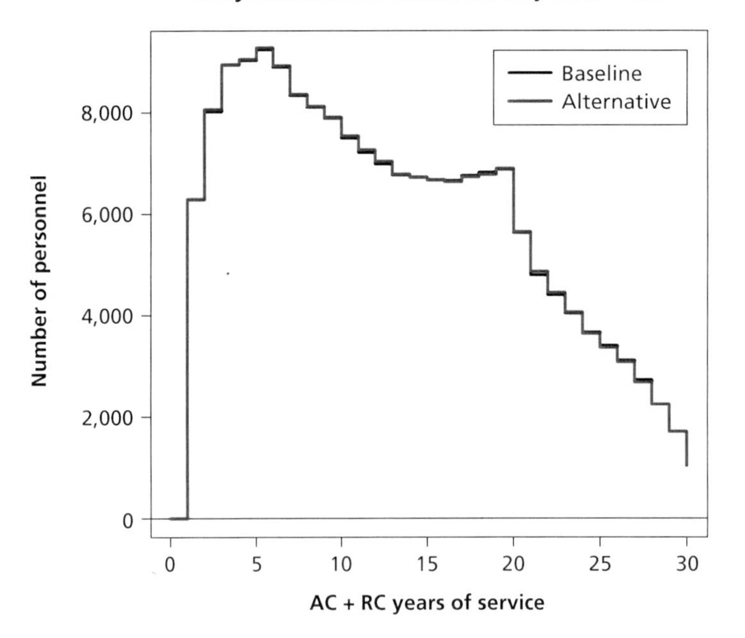

RAND *RR1022-A.1a*

Figure A.1b
Army AC Officer Retention and Prior-Active RC Officer Participation, 0-Percent DoD Match Rate, Annuity-Only Choice

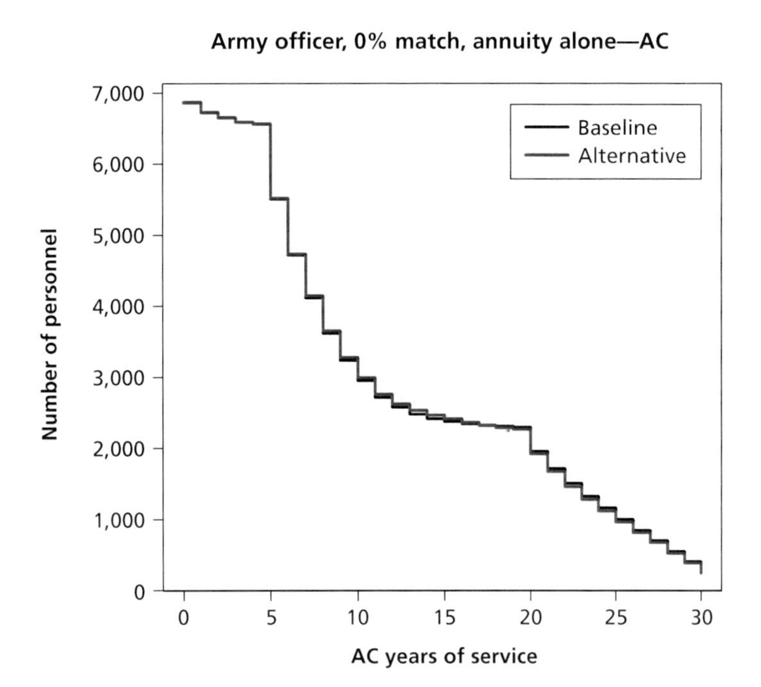

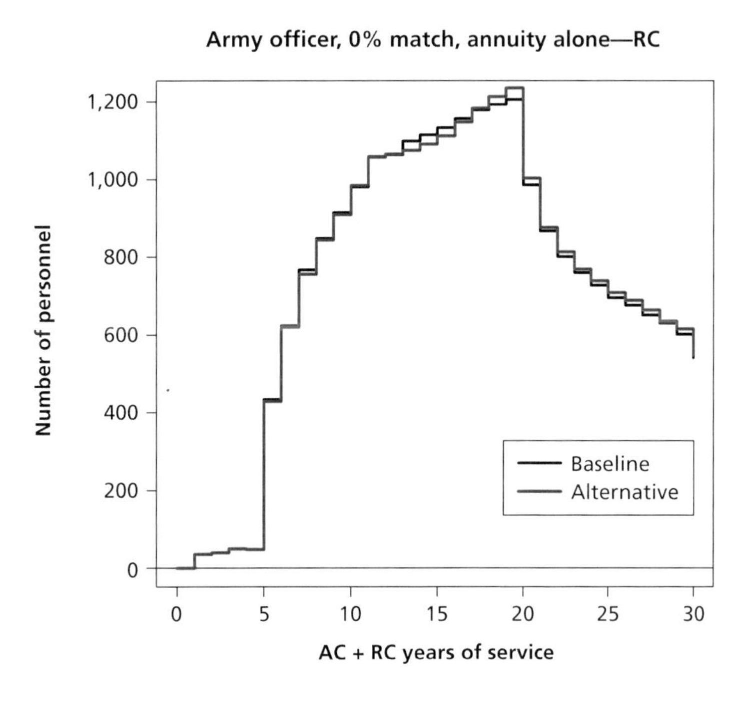

RAND *RR1022-A.1b*

Figure A.1c
Army AC Enlisted Retention and Prior-Active RC Enlisted Participation, 0-Percent DoD Match Rate, Partial Annuity/Partial Lump-Sum Choice

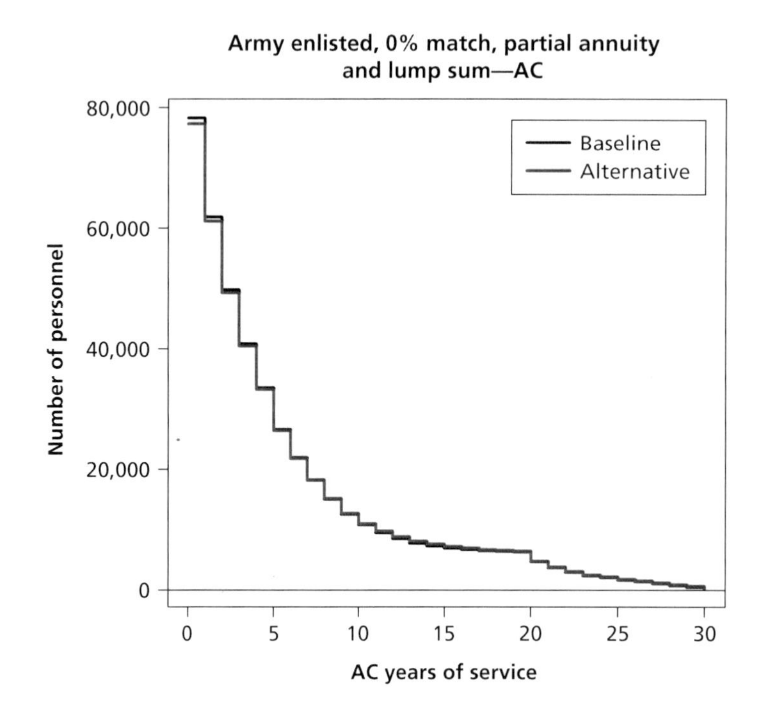

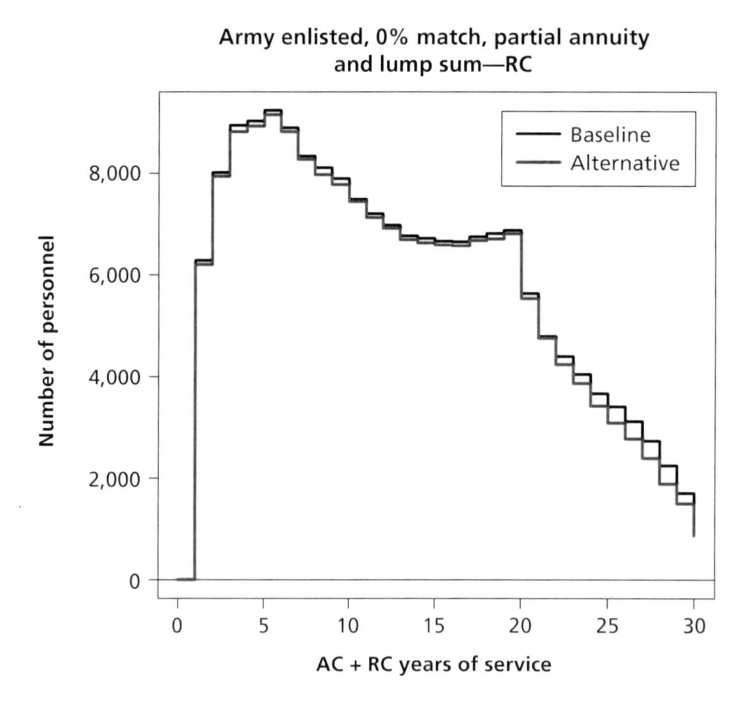

RAND *RR1022-A.1c*

Figure A.1d
Army AC Enlisted Retention and Prior-Active RC Enlisted Participation, 0-Percent DoD Match Rate, No Annuity/Full Lump-Sum Choice

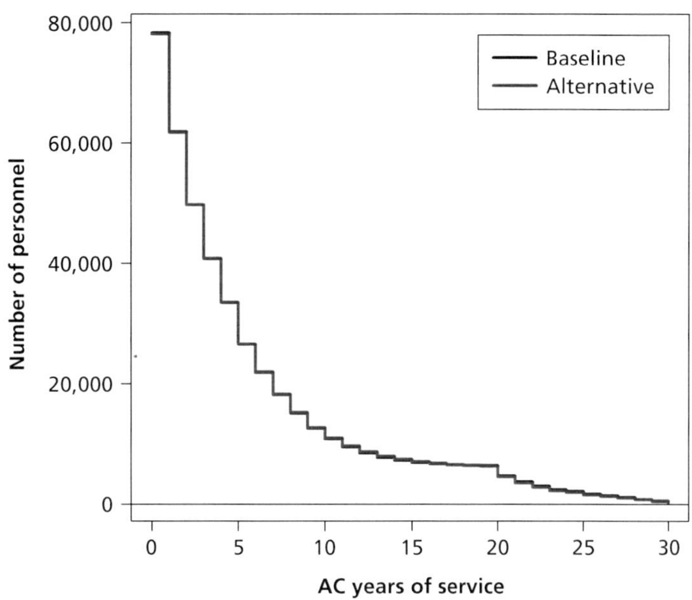

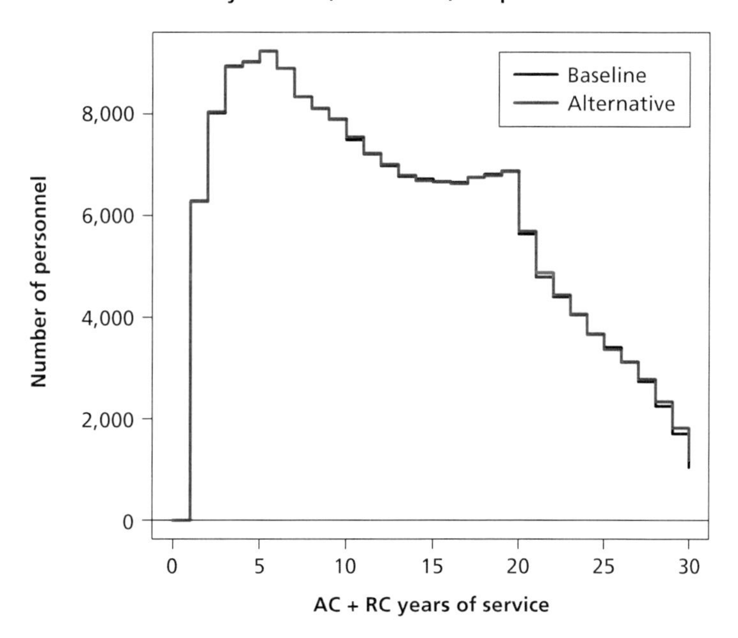

RAND *RR1022-A.1d*

Figure A.2a
Army AC Enlisted Retention and Prior-Active RC Enlisted Participation, 5-Percent DoD Match Rate, Annuity-Only Choice

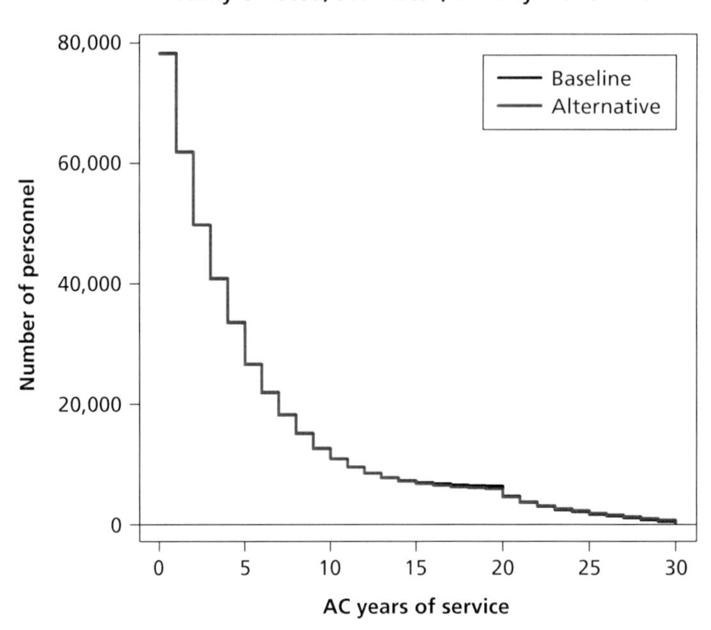

Army enlisted, 5% match, annuity alone—RC

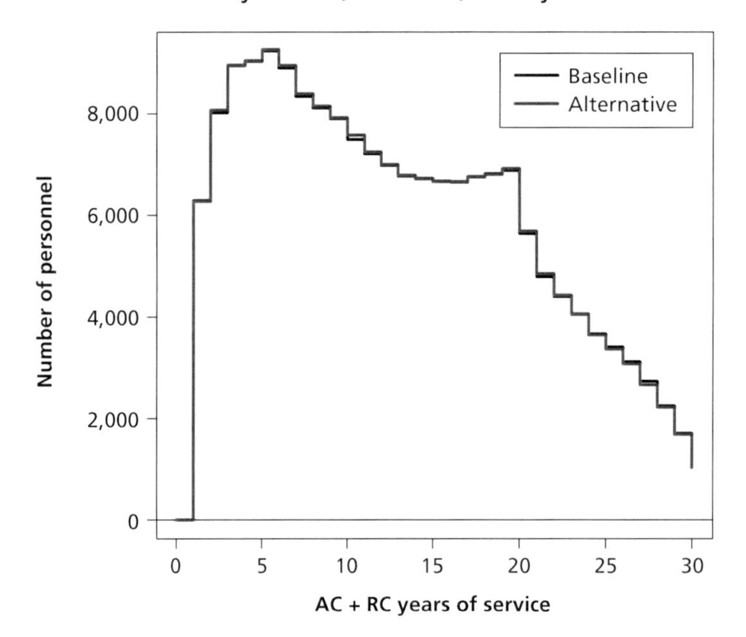

RAND *RR1022-A.2a*

Figure A.2b
Army AC Officer Retention and Prior-Active RC Officer Participation, 5-Percent DoD Match Rate, Annuity-Only Choice

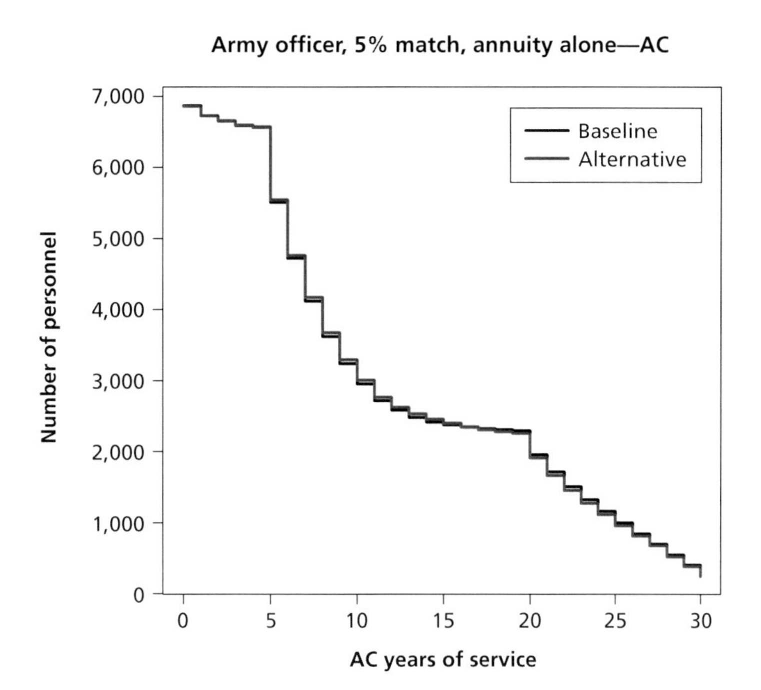

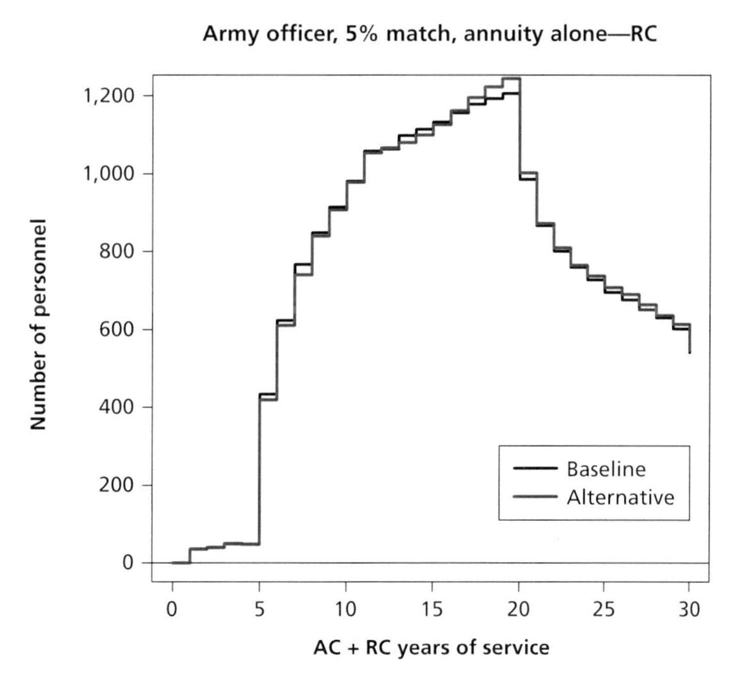

RAND *RR1022-A.2b*

Figure A.2c
Army AC Enlisted Retention and Prior-Active RC Enlisted Participation, 5-Percent DoD Match Rate, Partial Annuity/Partial Lump-Sum Choice

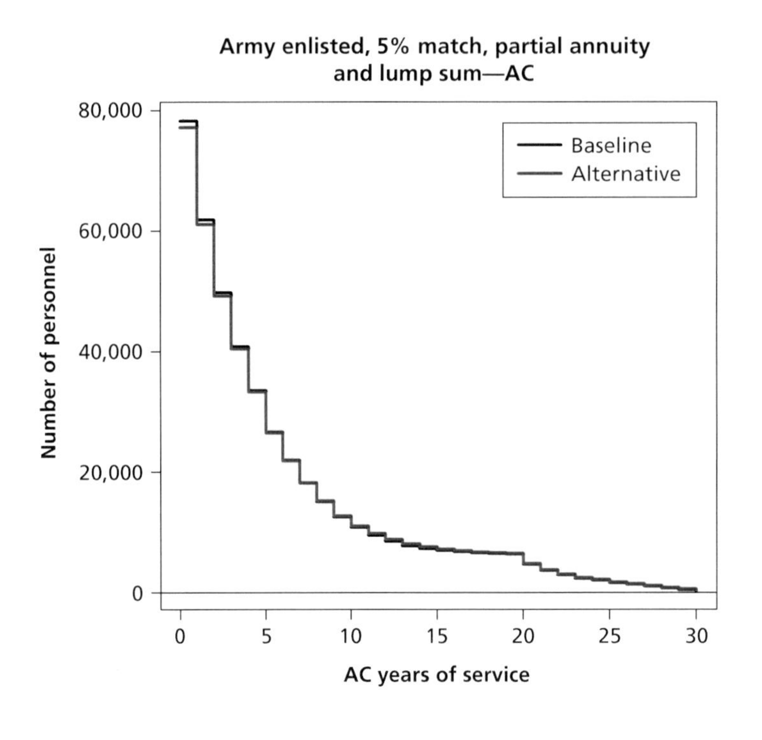

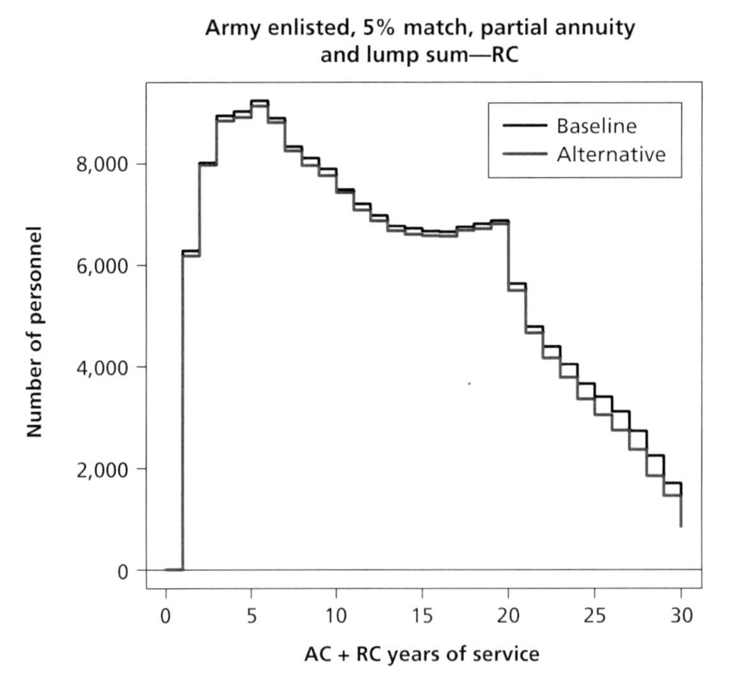

RAND *RR1022-A.2c*

Figure A.2d
Army AC Enlisted Retention and Prior-Active RC Enlisted Participation, 5-Percent DoD Match Rate, No Annuity/Full Lump-Sum Choice

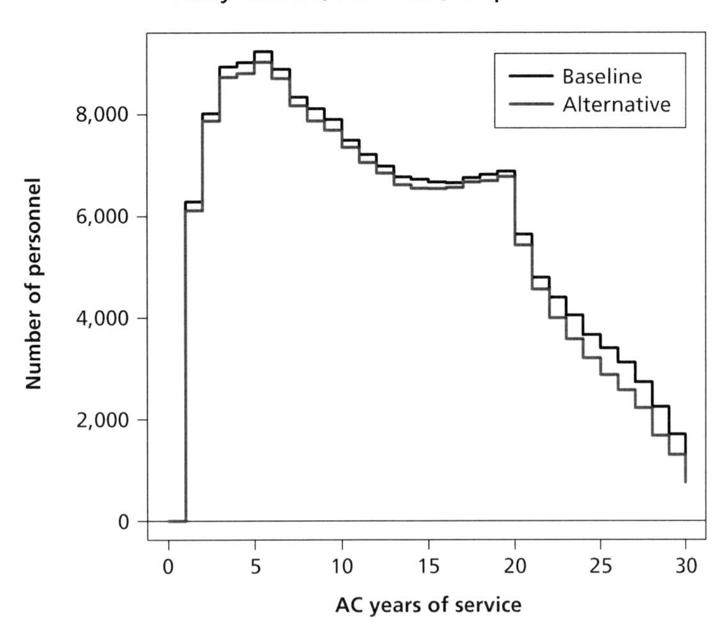

Army enlisted, 3% match, lump sum alone—RC

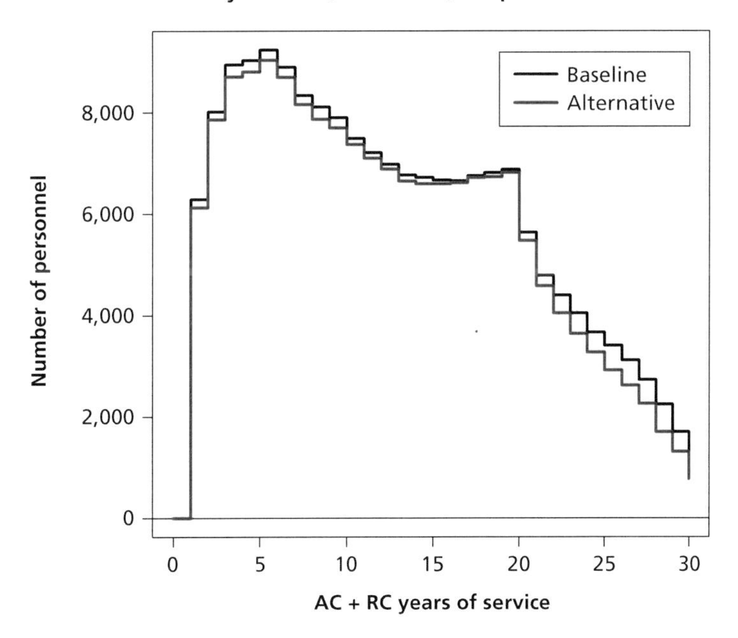

RAND *RR1022-A.2d*

Navy

Figure A.3a
Navy AC Enlisted Retention and Prior-Active RC Enlisted Participation, 0-Percent DoD Match Rate, Annuity Only Choice

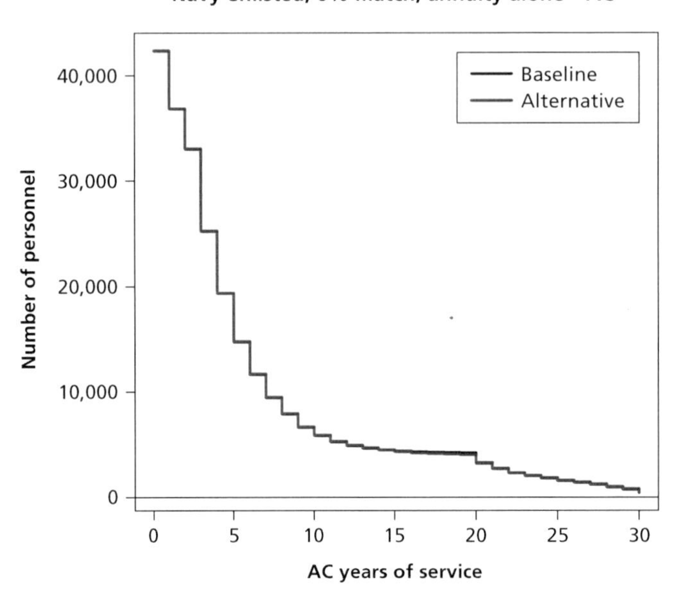

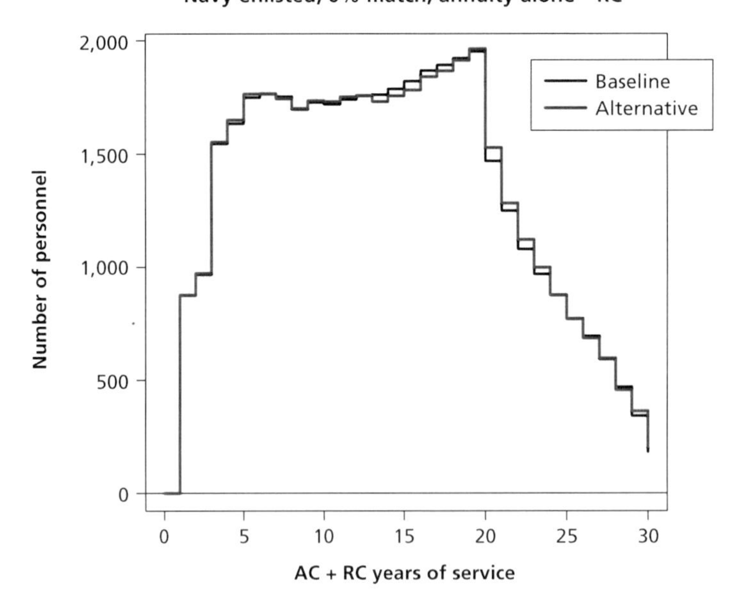

RAND *RR1022-A.3a*

Figure A.3b
Navy AC Officer Retention and Prior-Active RC Officer Participation, 0-Percent DoD Match Rate, Annuity-Only Choice

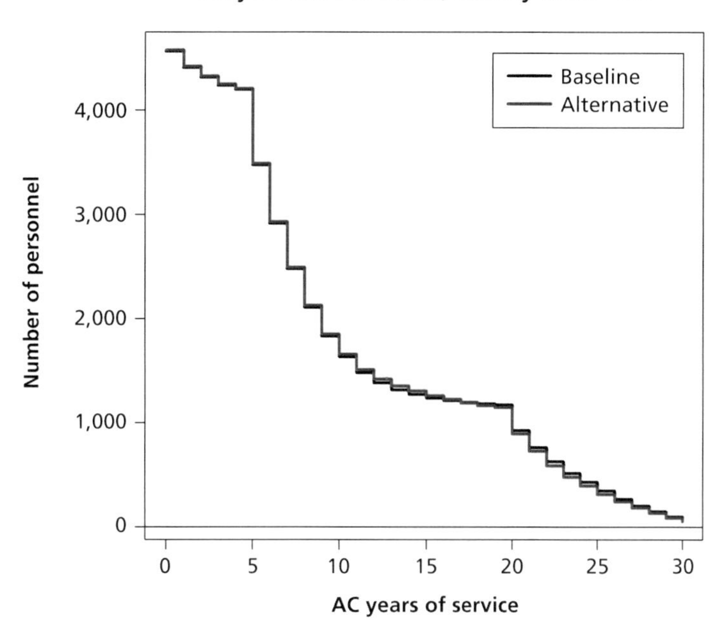

Navy officer, 0% match, annuity alone—RC

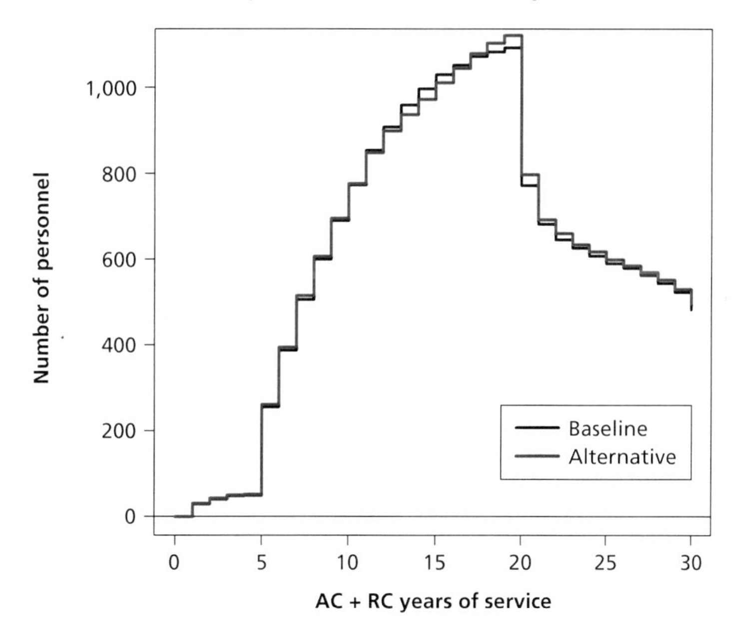

RAND *RR1022-A.3b*

Figure A.3c
Navy AC Enlisted Retention and Prior-Active RC Enlisted Participation, 0-Percent DoD Match Rate, Partial Annuity/Partial Lump-Sum Choice

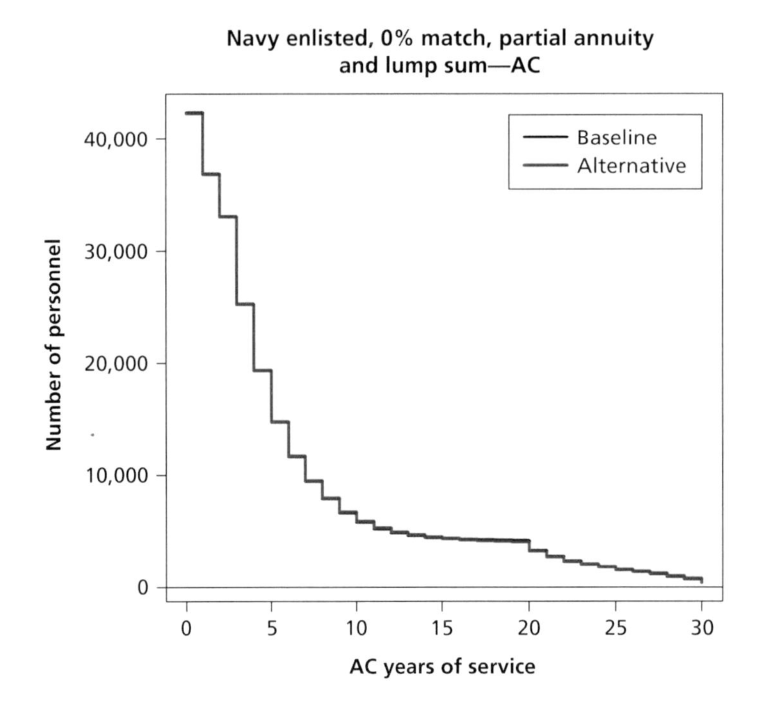

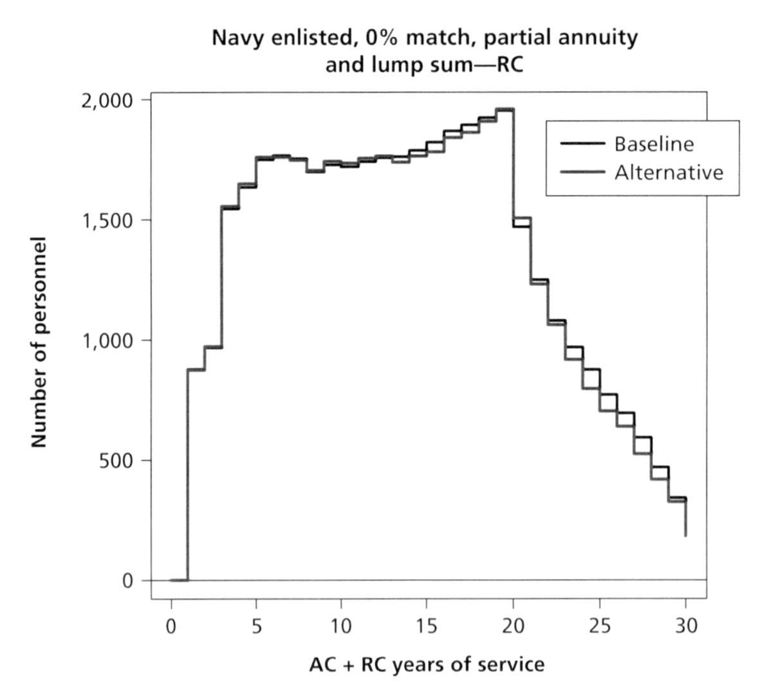

RAND *RR1022-A.3c*

Figure A.3d
Navy AC Enlisted Retention and Prior-Active RC Enlisted Participation, 0-Percent DoD Match Rate, No Annuity/Full Lump-Sum Choice

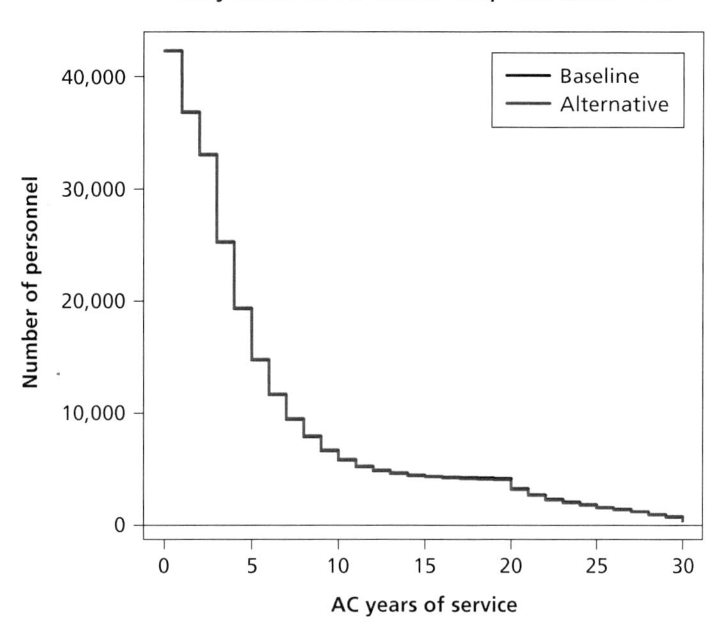

Navy enlisted, 0% match, lump sum alone—RC

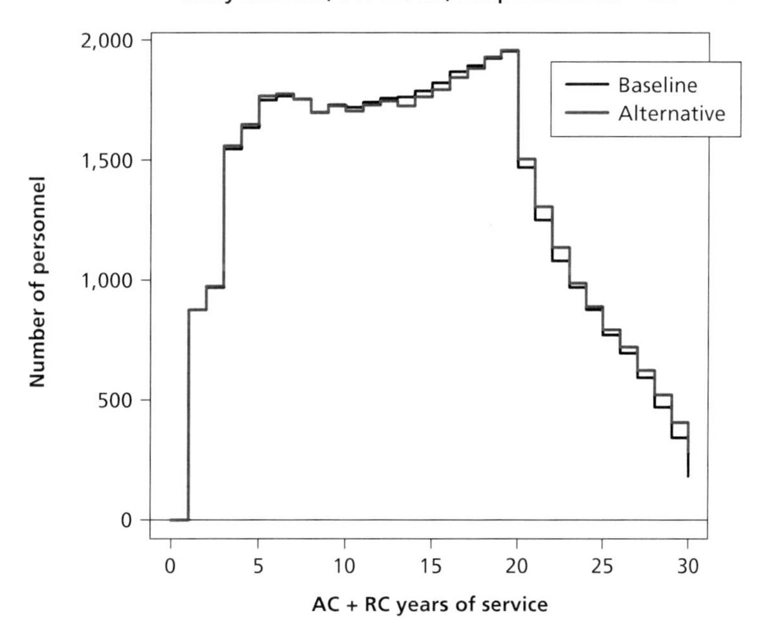

RAND *RR1022-A.3d*

Figure A.4a
Navy AC Enlisted Retention and Prior-Active RC Enlisted Participation, 3-Percent DoD Match Rate, Annuity-Only Choice

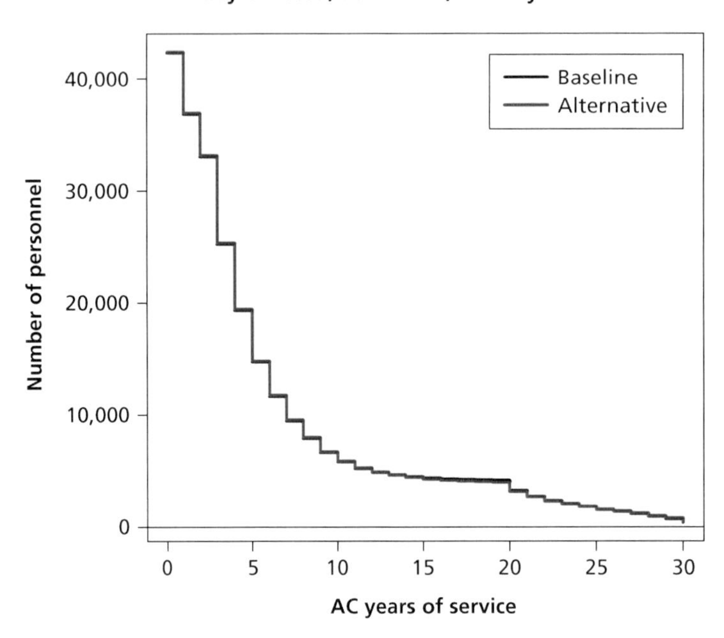

Navy enlisted, 3% match, annuity alone—RC

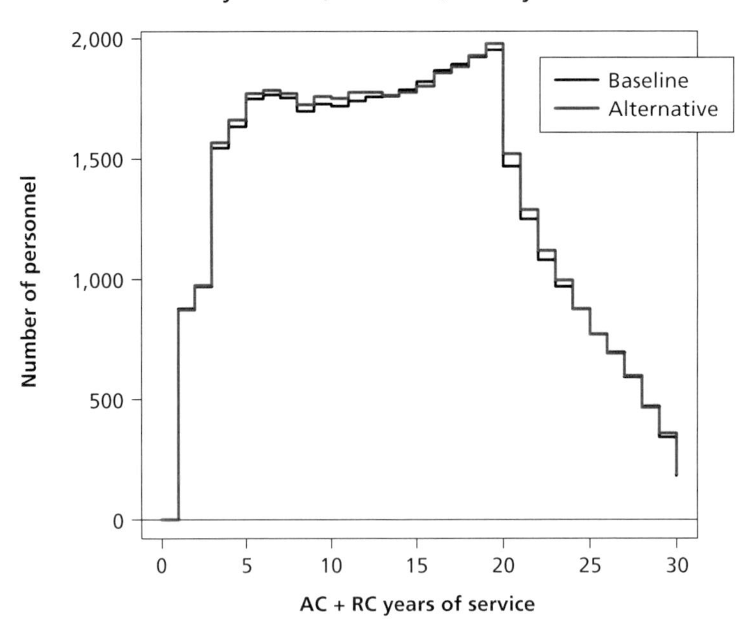

RAND *RR1022-A.4a*

Figure A.4b
Navy AC Officer Retention and Prior-Active RC Officer Participation, 3-Percent DoD Match Rate, Annuity-Only Choice

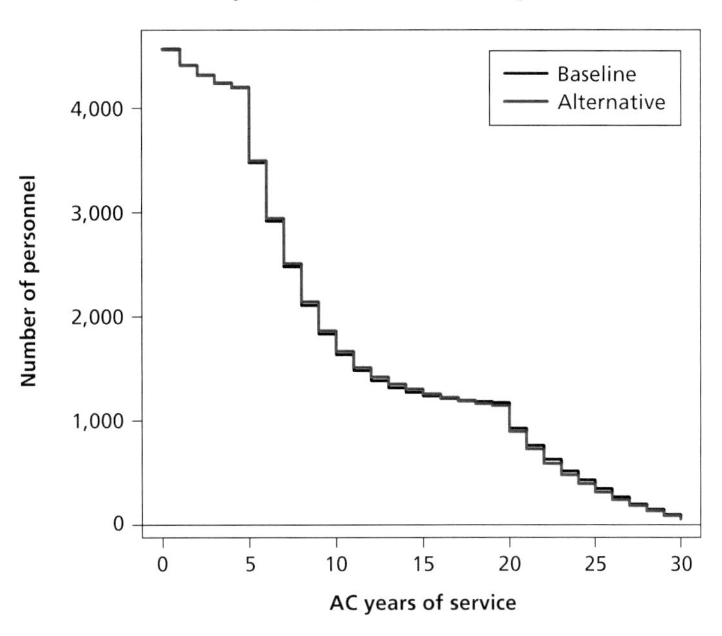

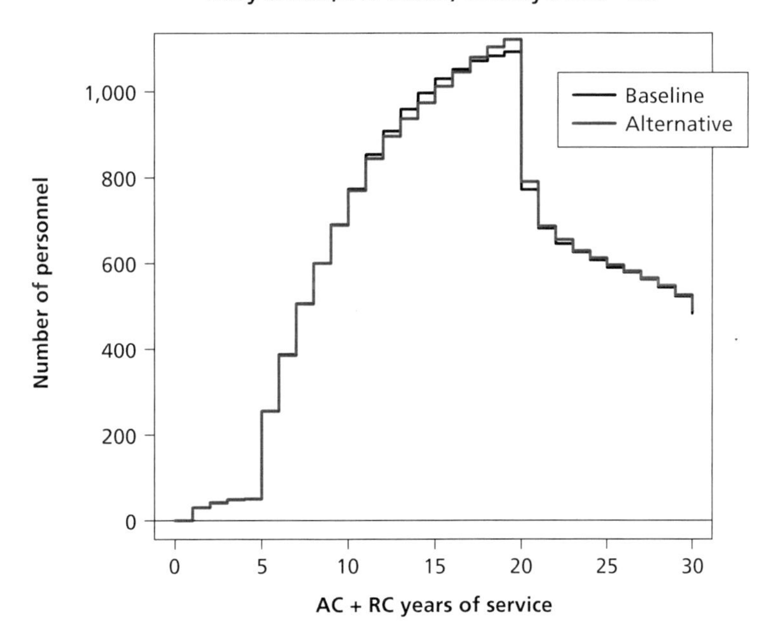

RAND *RR1022-A.4b*

Figure A.4c
Navy AC Enlisted Retention and Prior-Active RC Enlisted Participation, 3-Percent DoD Match Rate, Partial Annuity/Partial Lump-Sum Choice

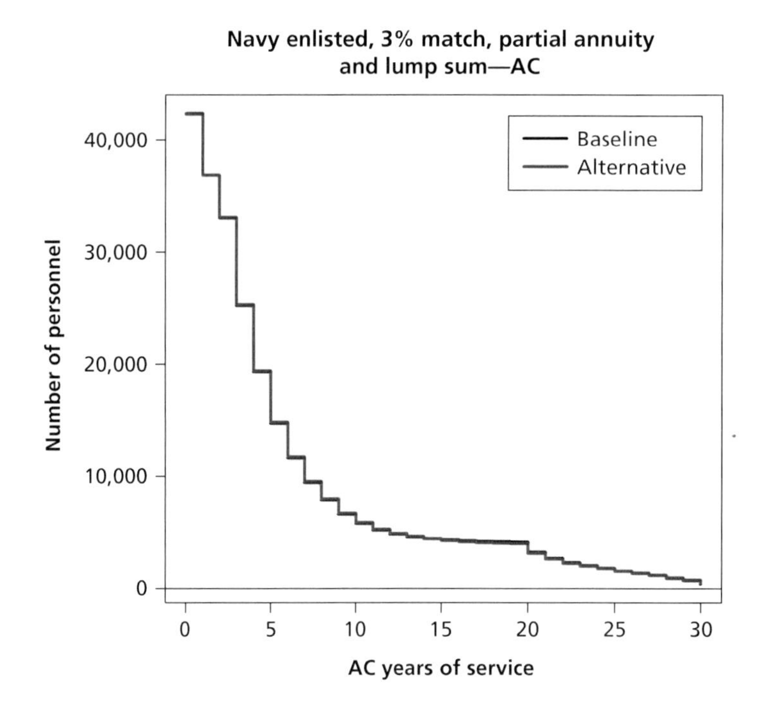

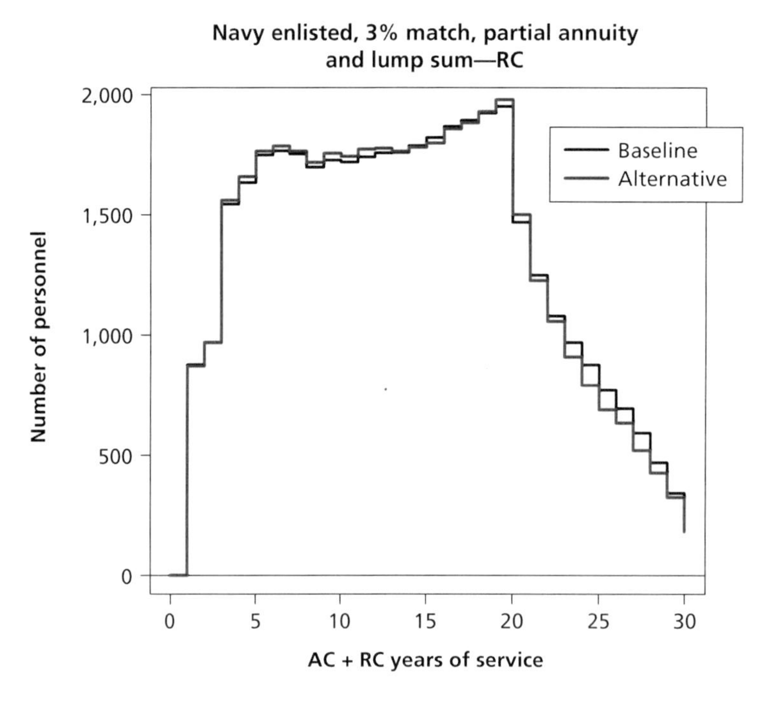

Figure A.4d
Navy AC Enlisted Retention and Prior-Active RC Enlisted Participation, 3-Percent DoD Match Rate, No Annuity/Full Lump-Sum Choice

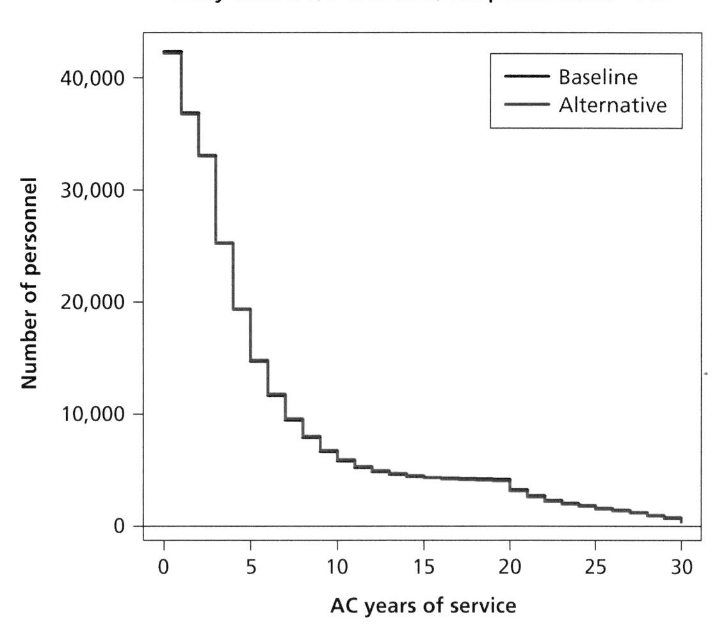

Navy enlisted, 3% match, lump sum alone—RC

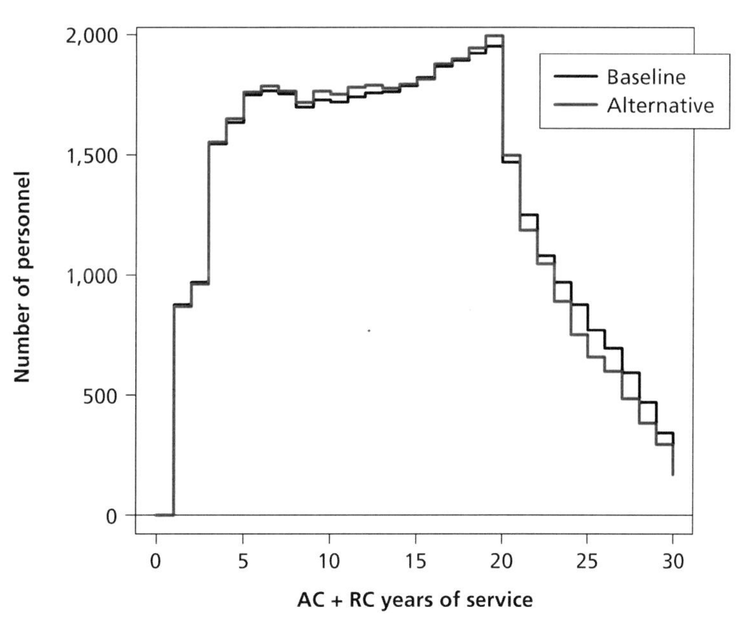

RAND *RR1022-A.4d*

Figure A.5a
Navy AC Enlisted Retention and Prior-Active RC Enlisted Participation, 5-Percent DoD Match Rate, Annuity-Only Choice

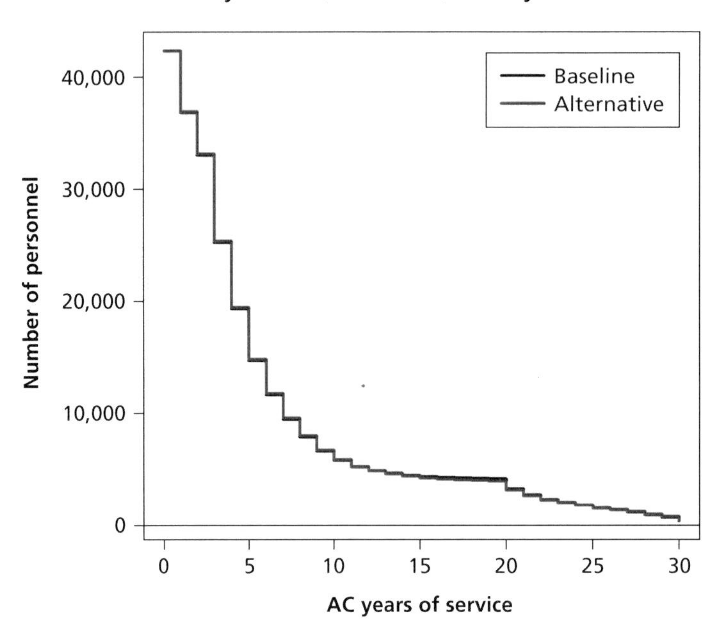

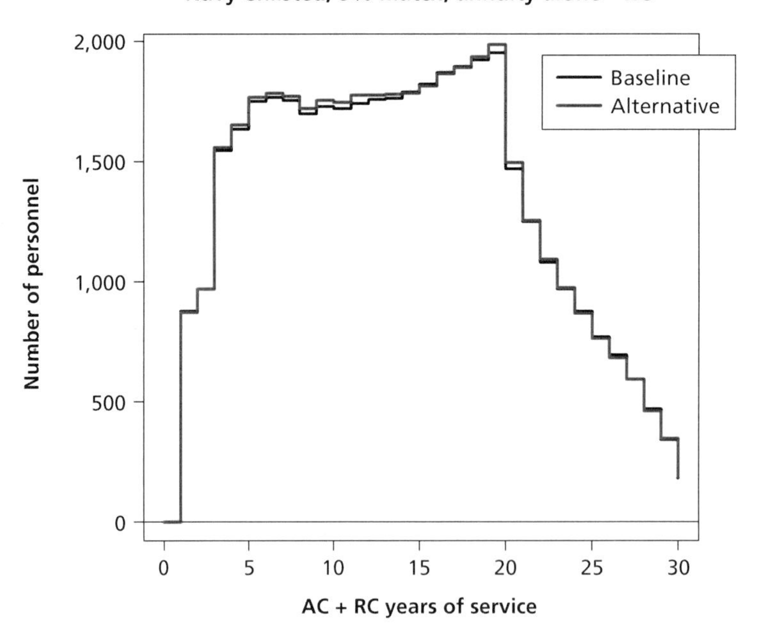

RAND *RR1022-A.5a*

Figure A.5b
Navy AC Officer Retention and Prior-Active RC Officer Participation, 5-Percent DoD Match Rate, Annuity-Only Choice

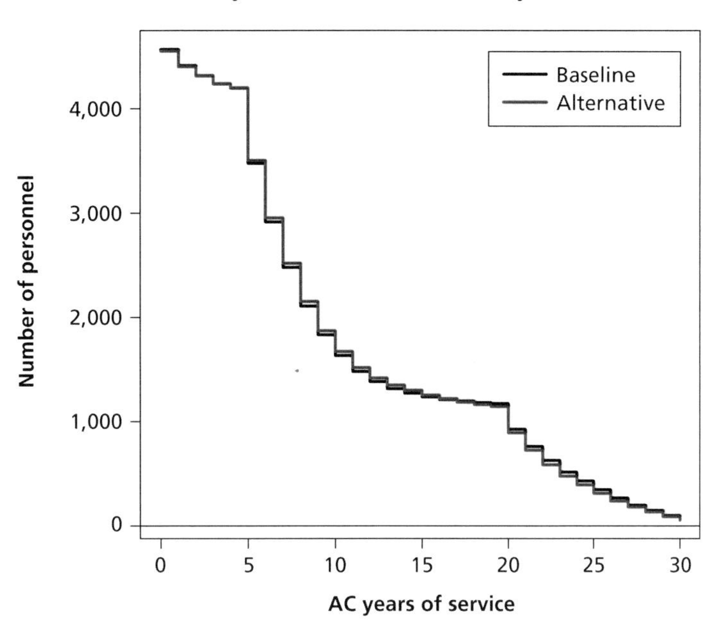

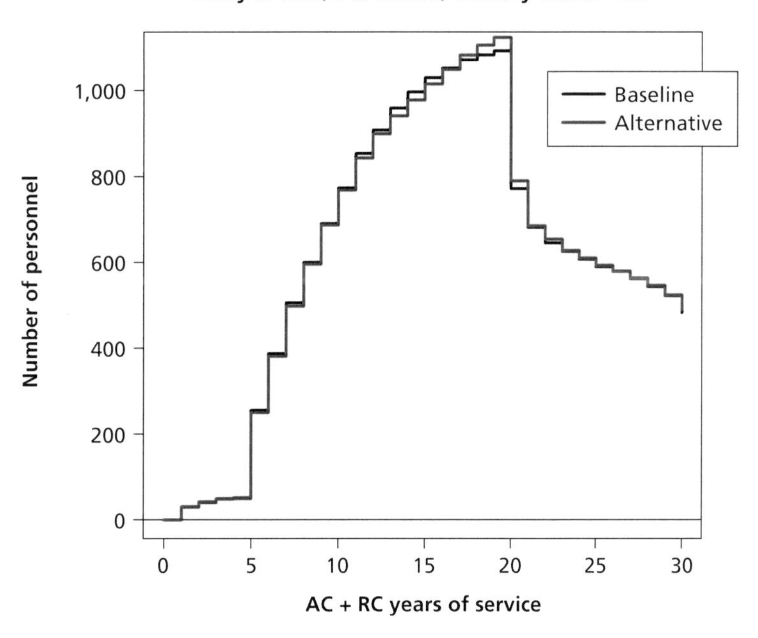

RAND *RR1022-A.5b*

Figure A.5c
Navy AC Enlisted Retention and Prior-Active RC Enlisted Participation, 5-Percent DoD Match Rate, Partial Annuity/Partial Lump-Sum Choice

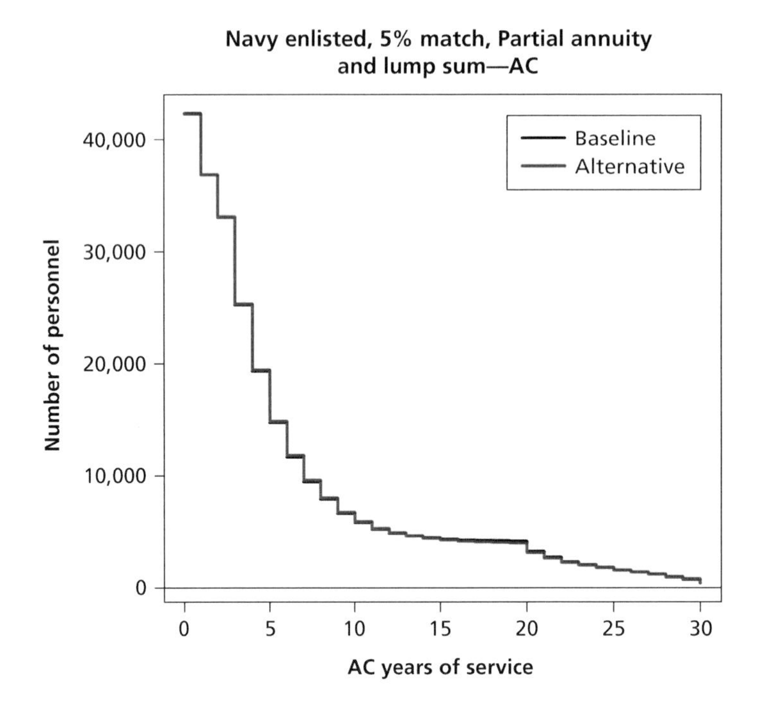

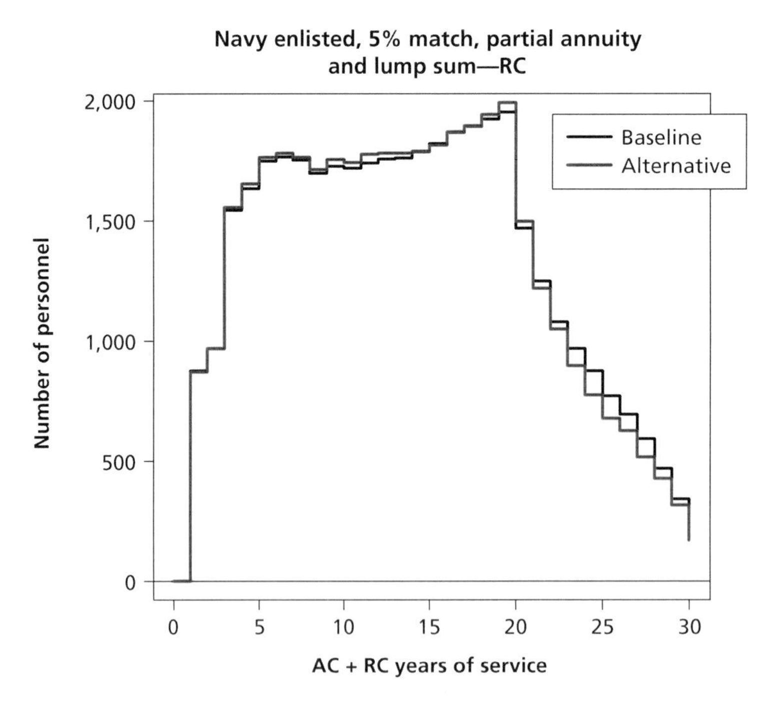

RAND *RR1022-A.5c*

Figure A.5d
Navy AC Enlisted Retention and Prior-Active RC Enlisted Participation, 5-Percent DoD Match Rate, No Annuity/Full Lump-Sum Choice

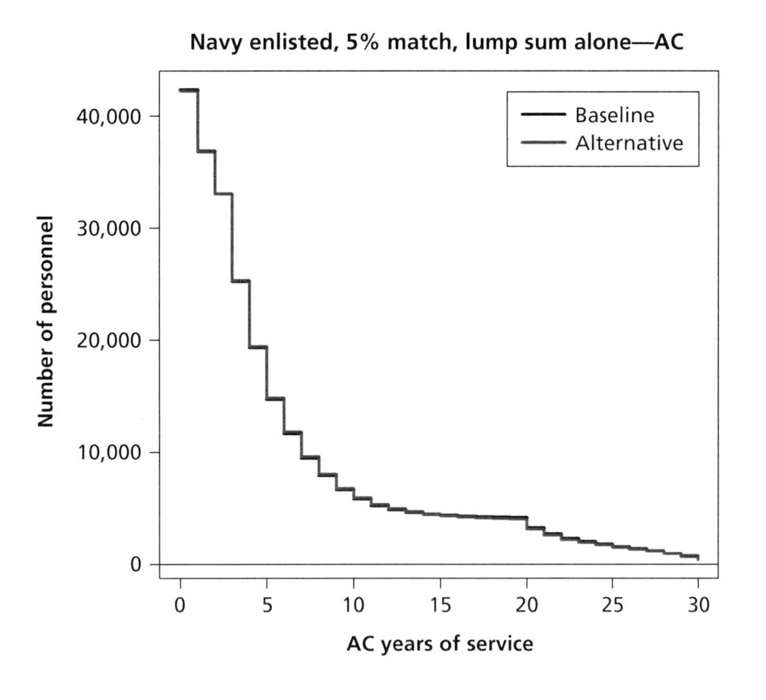

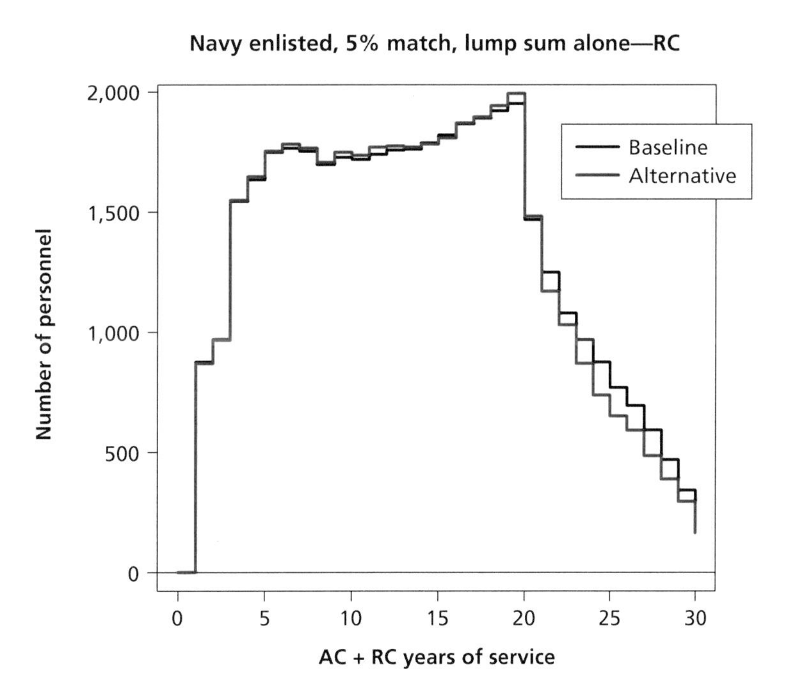

RAND *RR1022-A.5d*

Marine Corps

Figure A.6a
Marine Corps AC Enlisted Retention and Prior-Active RC Enlisted Participation, 0-Percent DoD Match Rate, Annuity-Only Choice

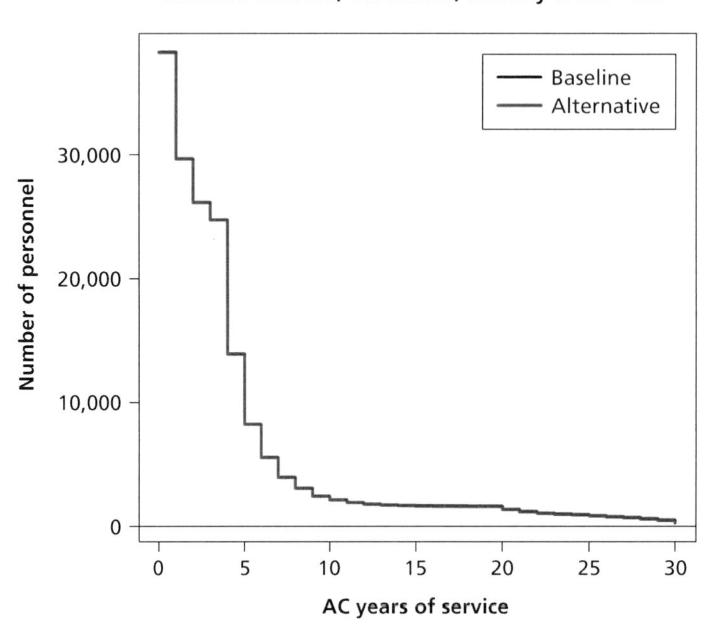

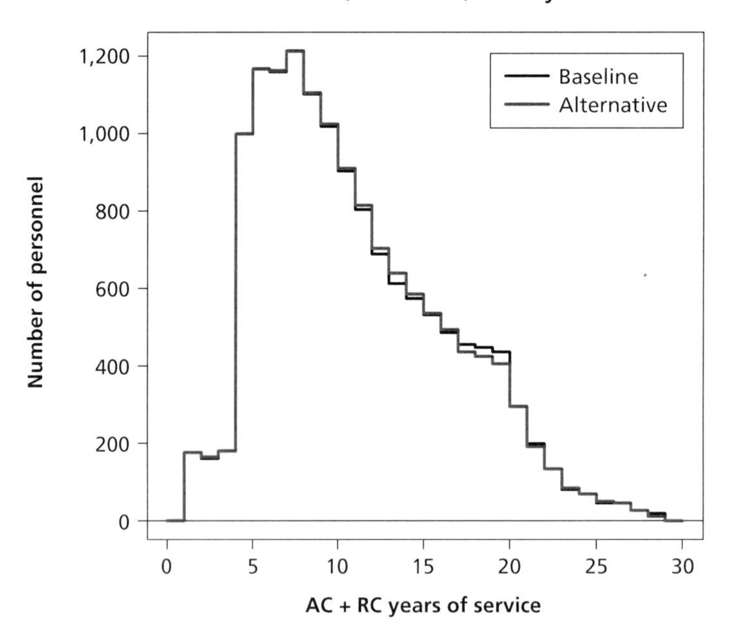

RAND *RR1022-A.6a*

Figure A.6b
Marine Corps AC Officer Retention and Prior-Active RC Officer Participation, 0-Percent DoD Match Rate, Annuity-Only Choice

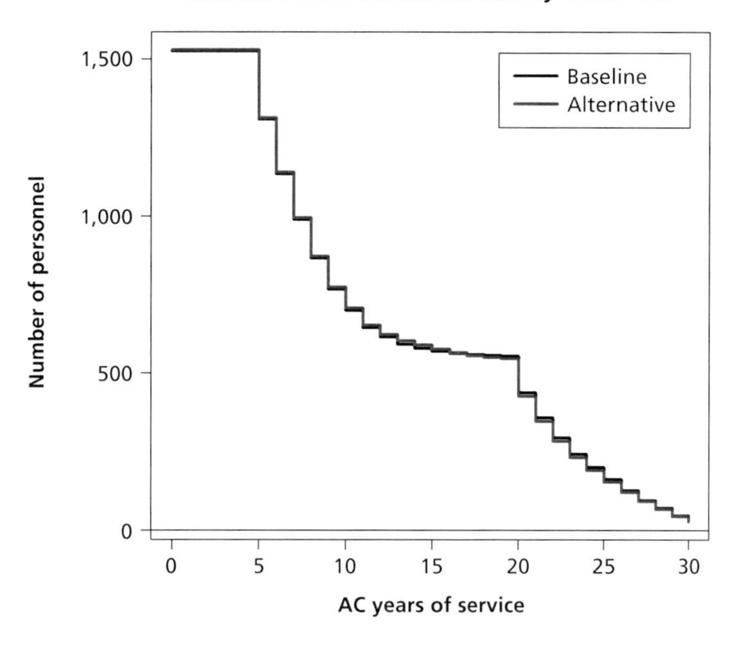

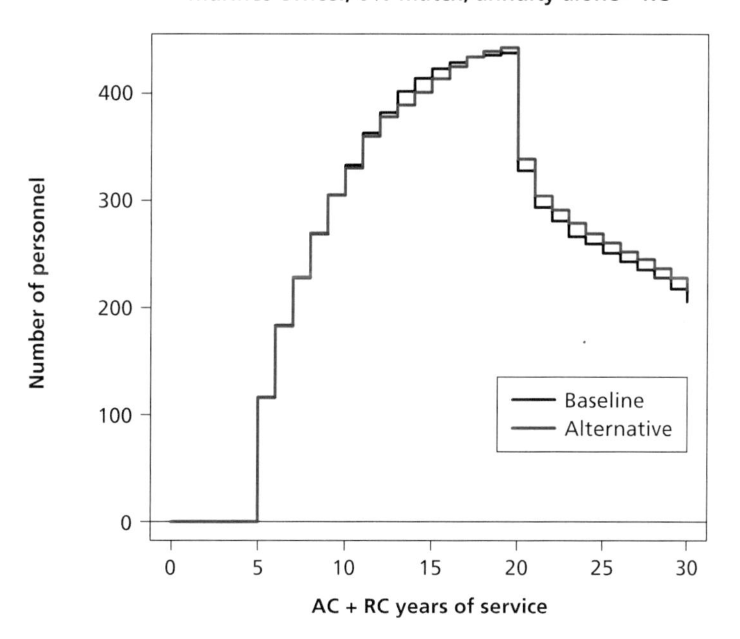

RAND *RR1022-A.6b*

Figure A.6c
Marine Corps AC Enlisted Retention and Prior-Active RC Enlisted Participation, 0-Percent DoD Match Rate, Partial Annuity/Partial Lump-Sum Choice

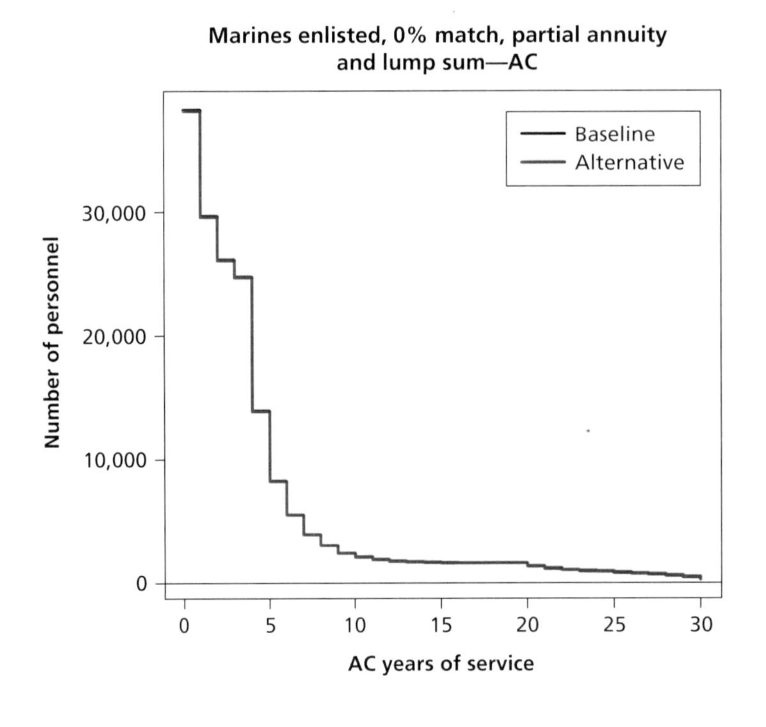

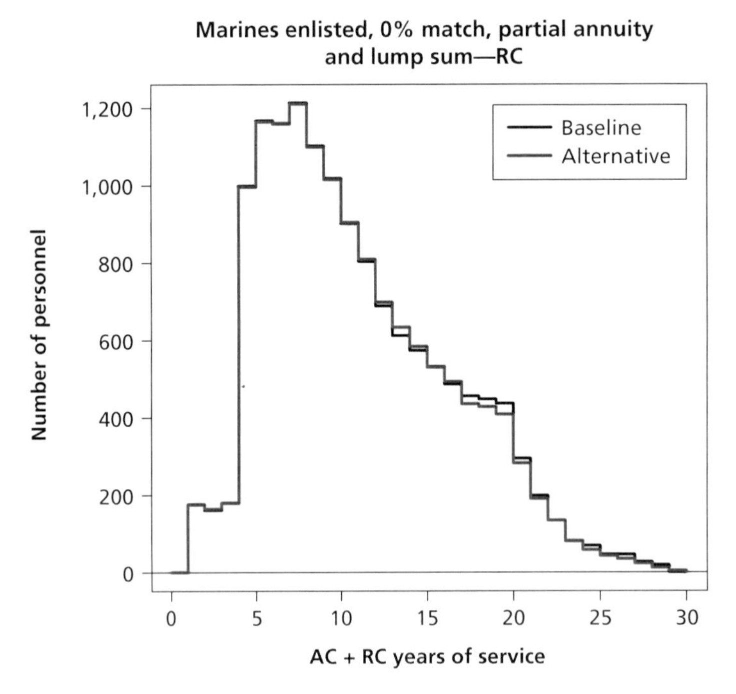

RAND *RR1022-A.6c*

Figure A.6d
Marine Corps AC Enlisted Retention and Prior-Active RC Enlisted Participation, 0-Percent DoD Match Rate, No Annuity/Full Lump-Sum Choice

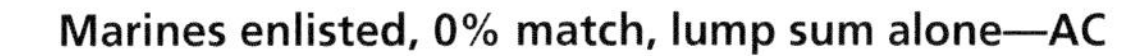

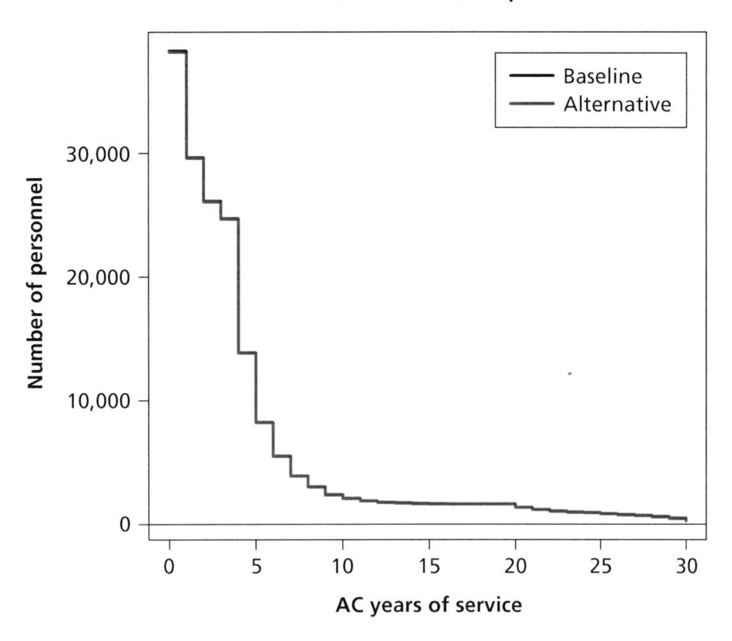

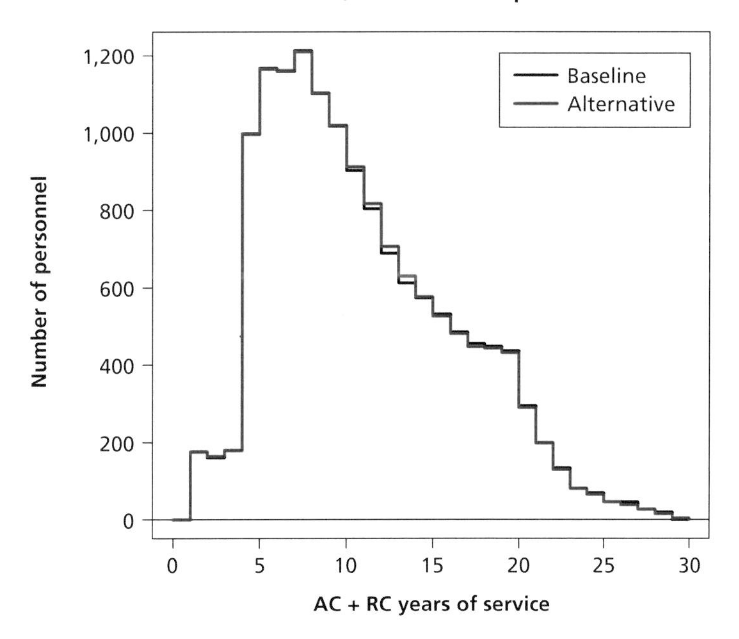

RAND *RR1022-A.6d*

Figure A.7a
Marine Corps AC Enlisted Retention and Prior-Active RC Enlisted Participation, 3-Percent DoD Match Rate, Annuity-Only Choice

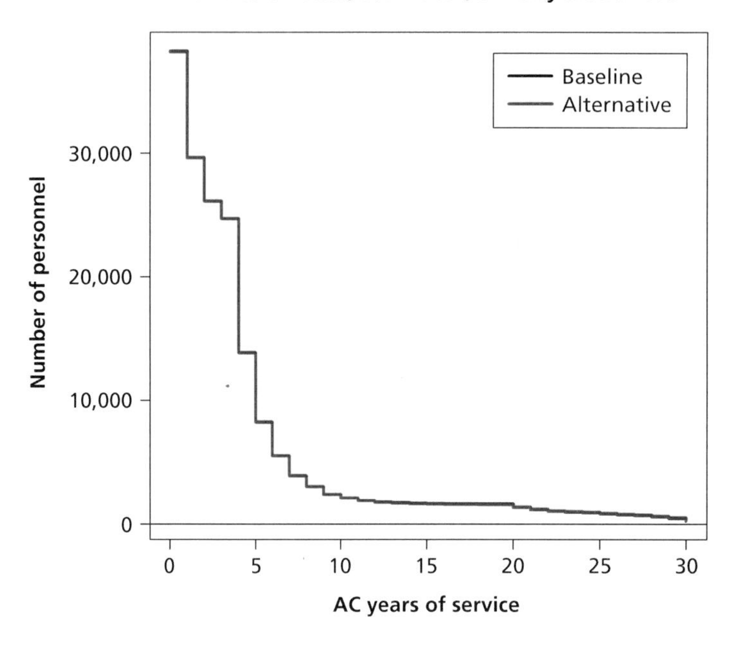

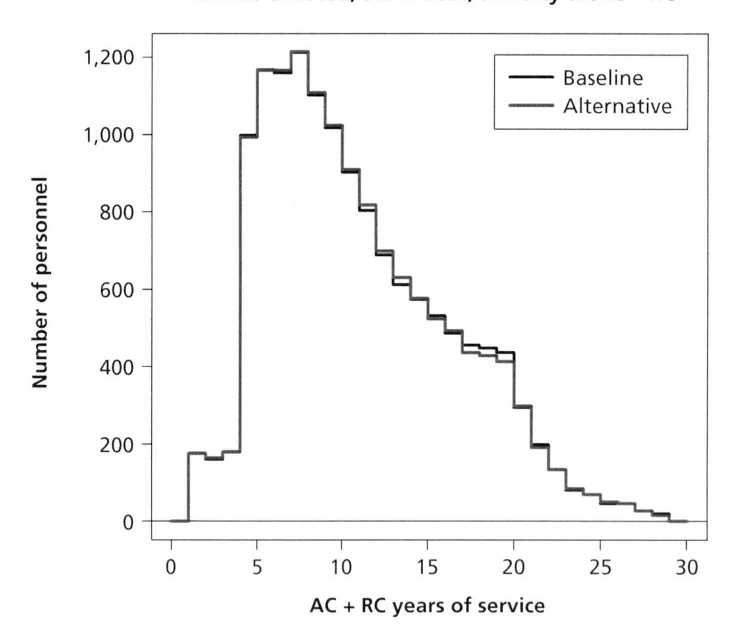

RAND *RR1022-A.7a*

Figure A.7b
Marine Corps AC Officer Retention and Prior-Active RC Officer Participation, 3-Percent DoD Match Rate, Annuity-Only Choice

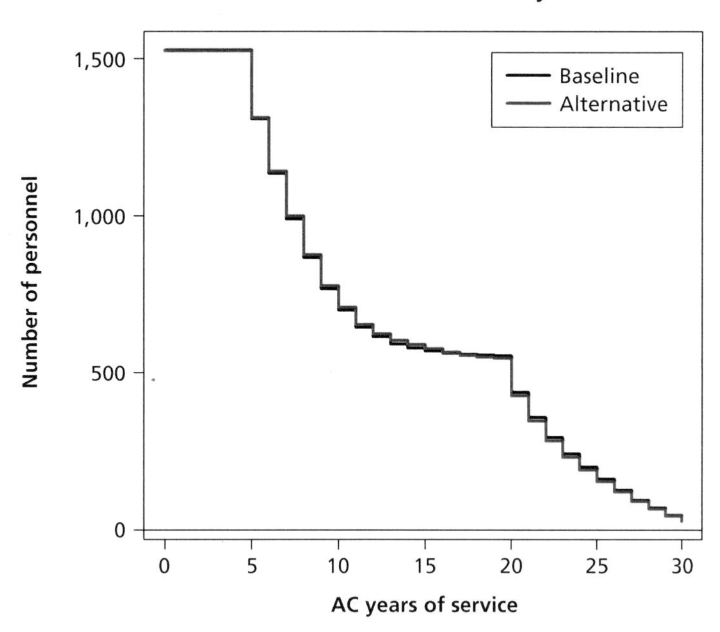

Marines officer, 3% match, annuity alone—RC

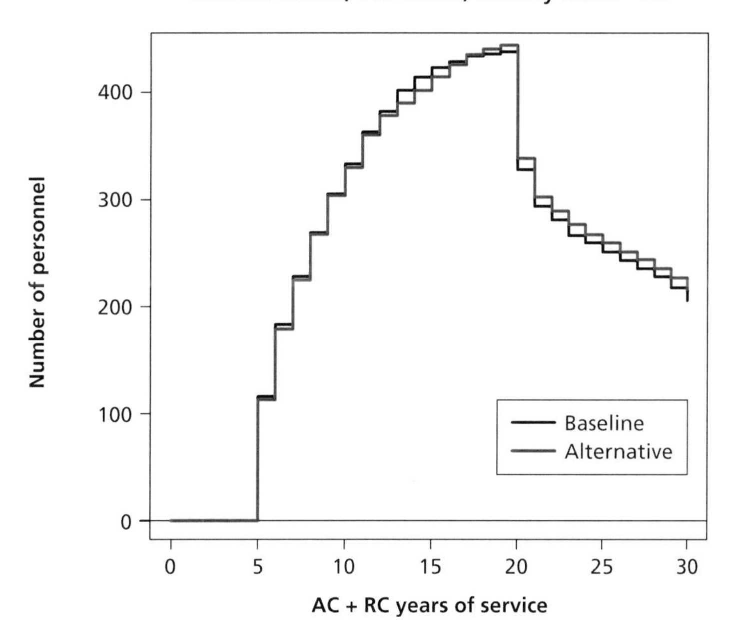

RAND *RR1022-A.7b*

Figure A.7c
Marine Corps AC Enlisted Retention and Prior-Active RC Enlisted Participation, 3-Percent DoD Match Rate, Partial Annuity/Partial Lump-Sum Choice

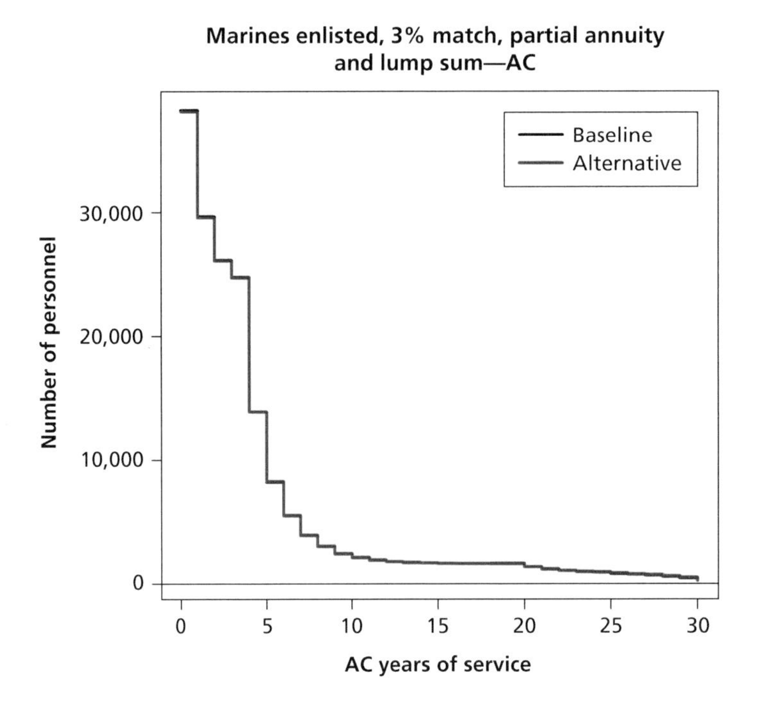

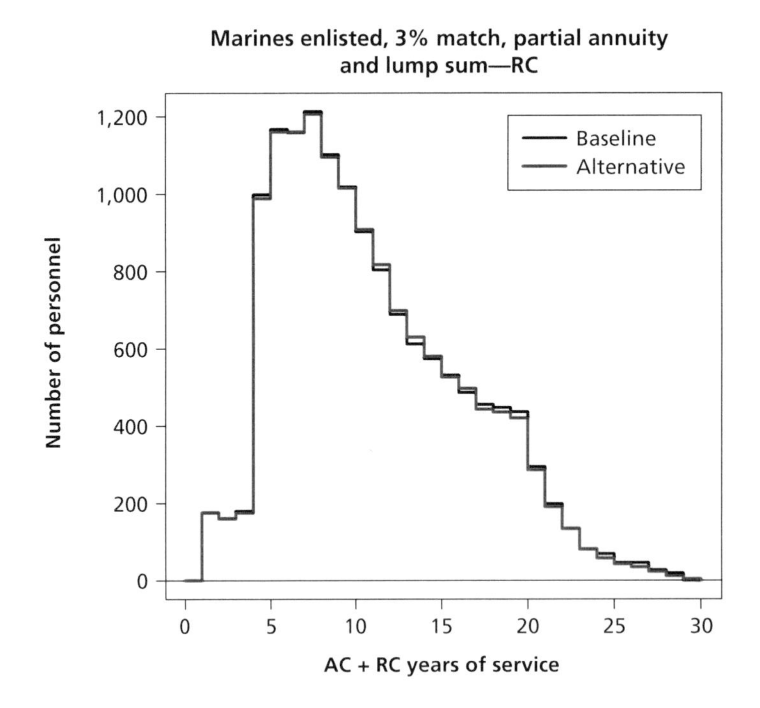

RAND *RR1022-A.7c*

Figure A.7d
Marine Corps AC Enlisted Retention and Prior-Active RC Enlisted Participation, 3-Percent DoD Match Rate, No Annuity/Full Lump-Sum Choice

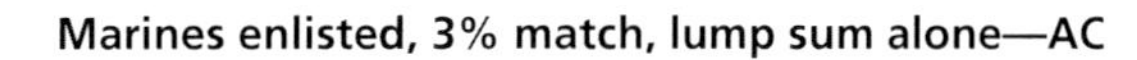

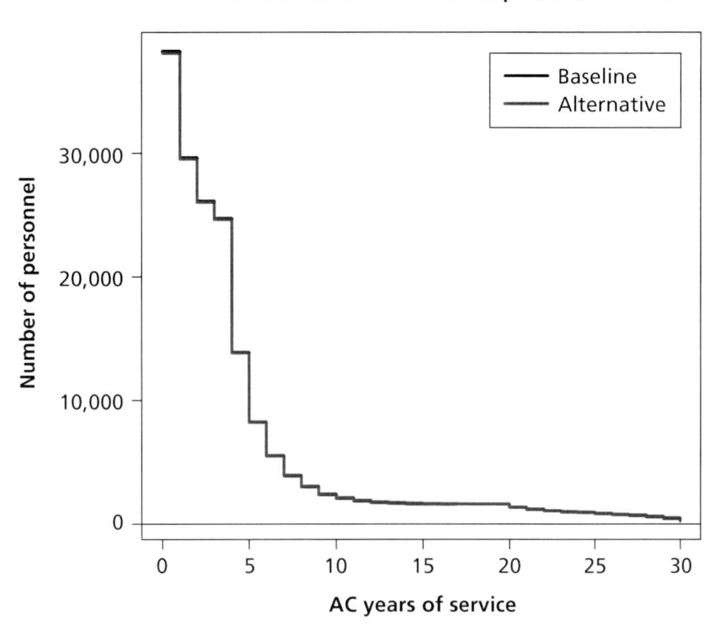

Marines enlisted, 3% match, lump sum alone—RC

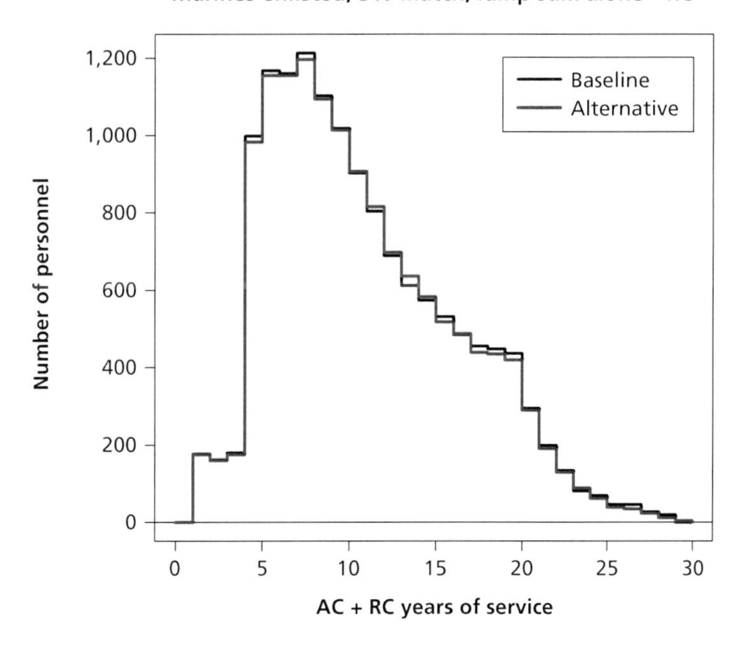

RAND *RR1022-A.7d*

Figure A.8a
Marine Corps AC Enlisted Retention and Prior-Active RC Enlisted Participation, 5-Percent DoD Match Rate, Annuity-Only Choice

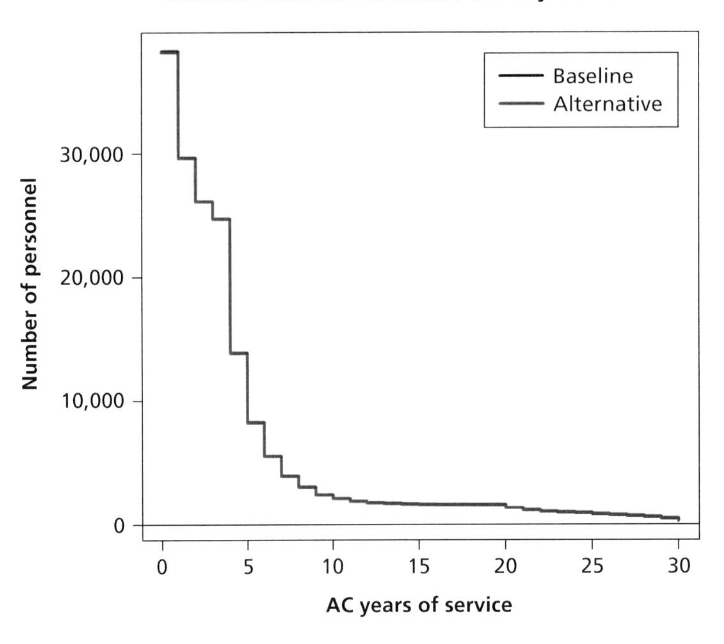

Marines enlisted, 5% match, annuity alone—RC

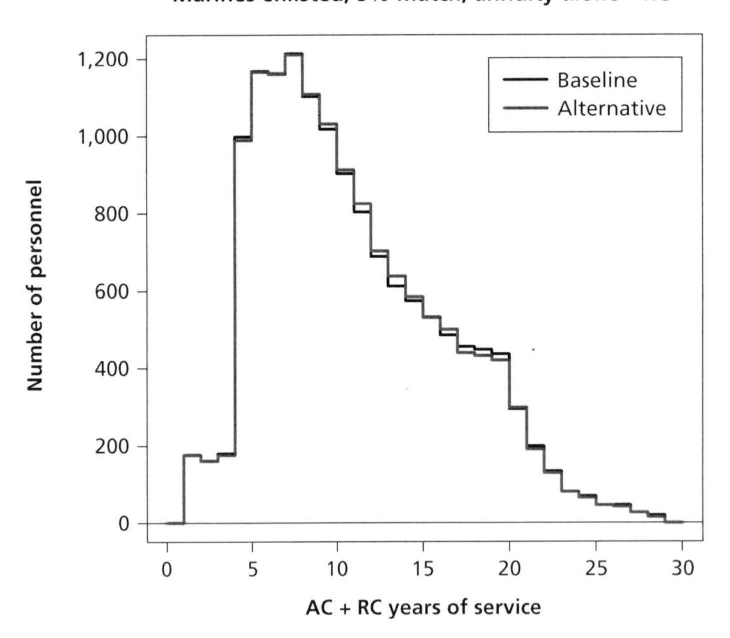

RAND *RR1022-A.8a*

Figure A.8b
Marine Corps AC Officer Retention and Prior-Active RC Officer Participation, 5-Percent DoD Match Rate, Annuity-Only Choice

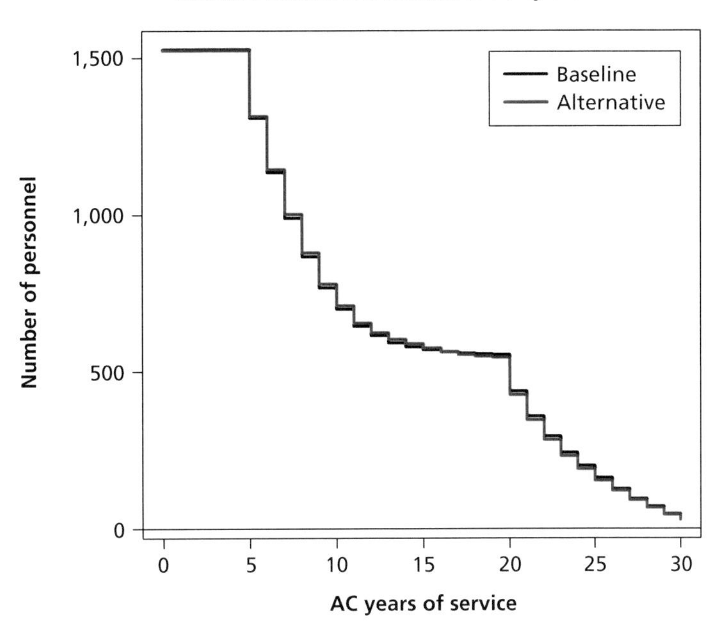

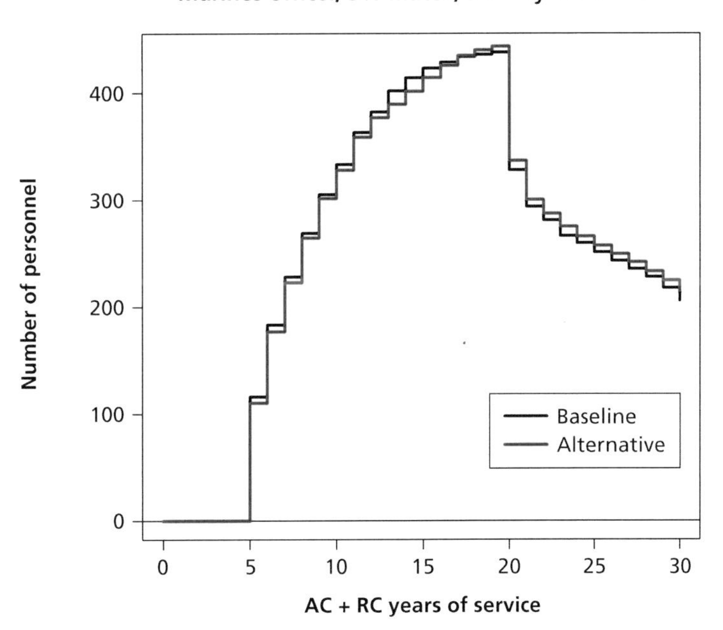

RAND *RR1022-A.8b*

Figure A.8c
Marine Corps AC Enlisted Retention and Prior-Active RC Enlisted Participation, 5-Percent DoD Match Rate, Partial Annuity/Partial Lump-Sum Choice

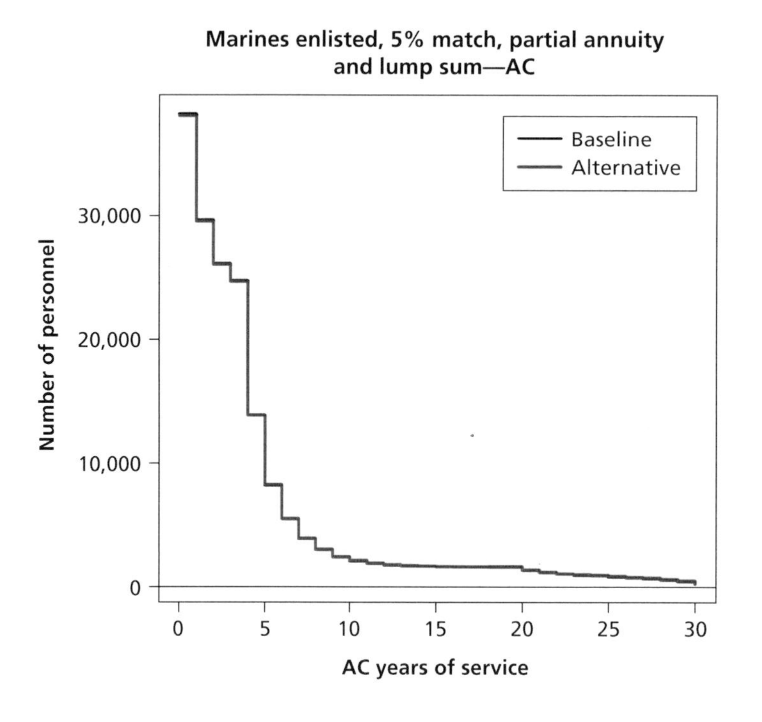

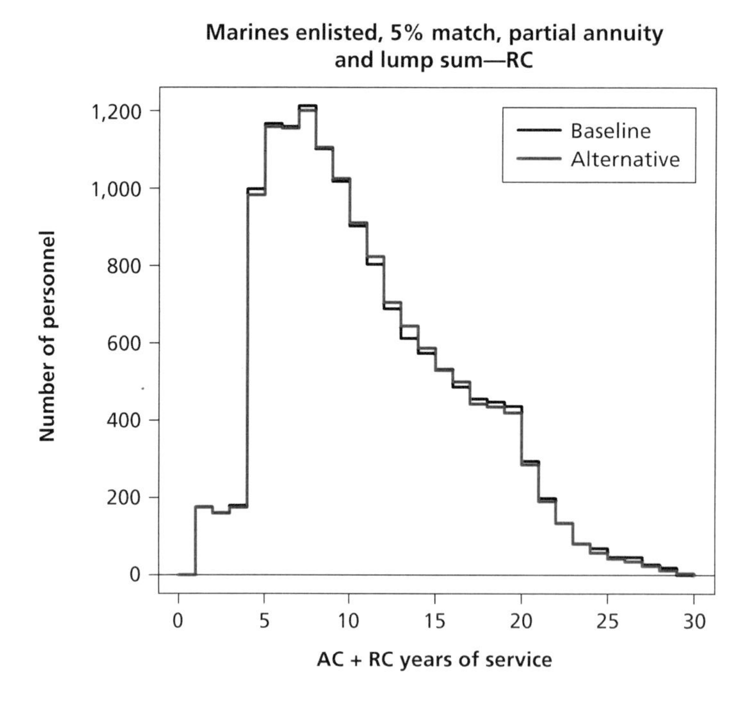

RAND *RR1022-A.8c*

Figure A.8d
Marine Corps AC Enlisted Retention and Prior-Active RC Enlisted Participation, 5-Percent DoD Match Rate, No Annuity/Full Lump-Sum Choice

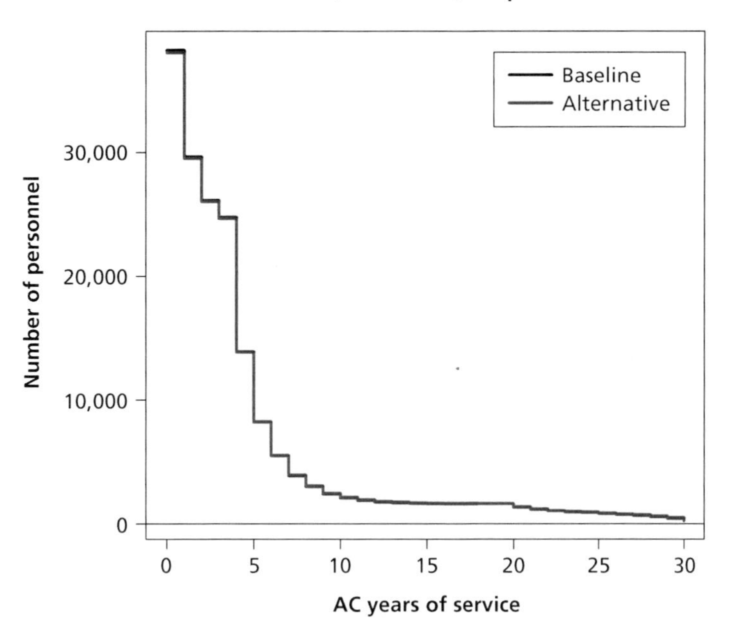

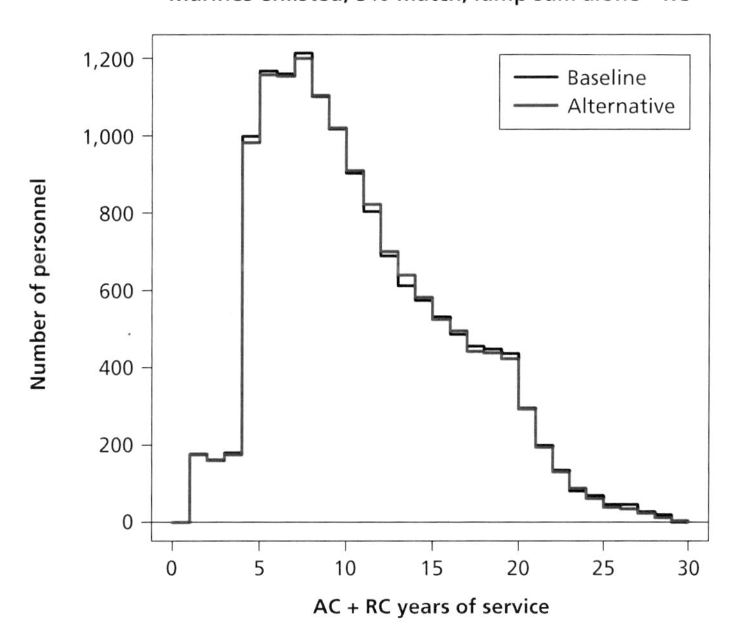

RAND *RR1022-A.8d*

Air Force

Figure A.9a
Air Force AC Enlisted Retention and Prior-Active RC Enlisted Participation, 0-Percent DoD Match Rate, Annuity-Only Choice

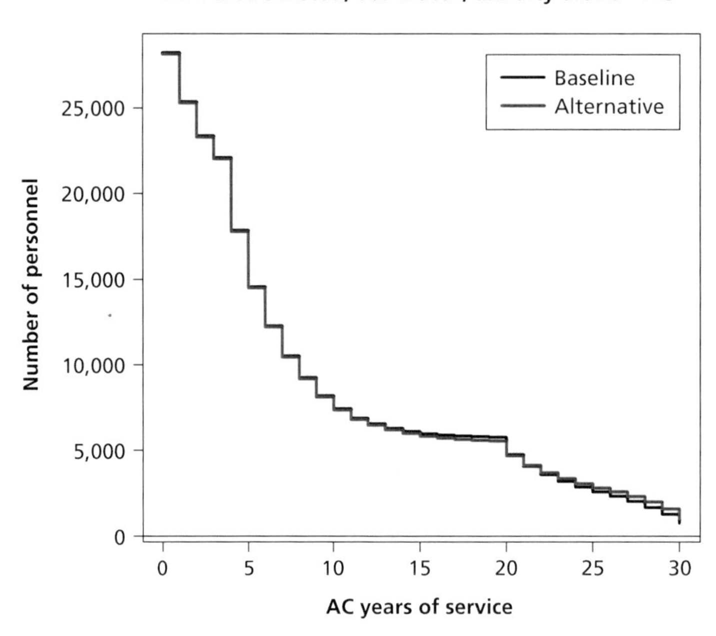

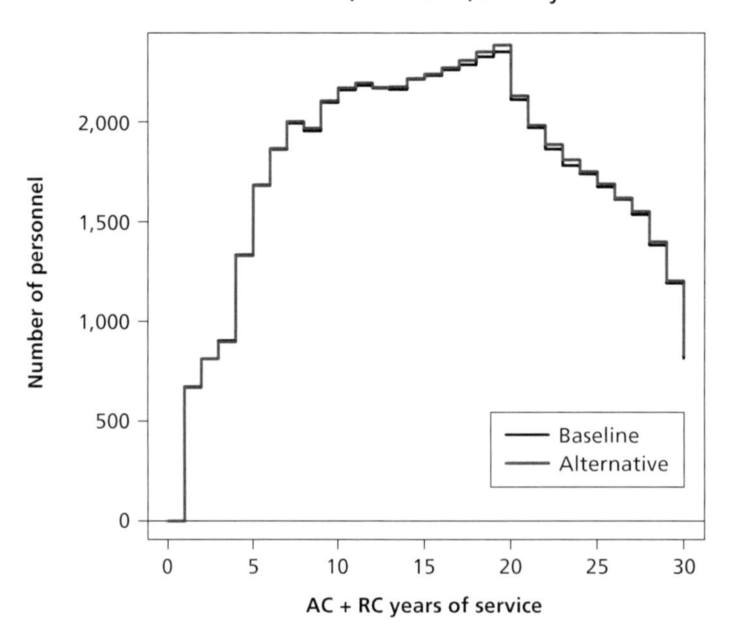

RAND *RR1022-A.9a*

Figure A.9b
Air Force AC Officer Retention and Prior-Active RC Officer Participation, 0-Percent DoD Match Rate, Annuity-Only Choice

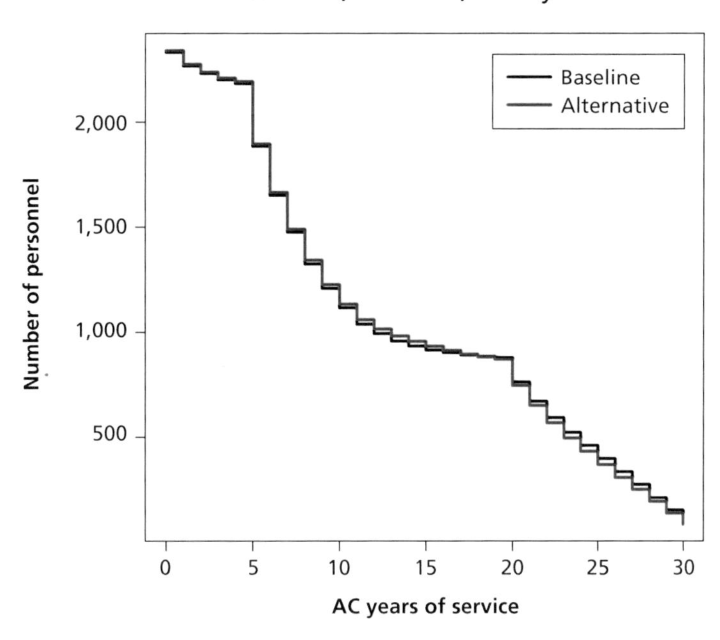

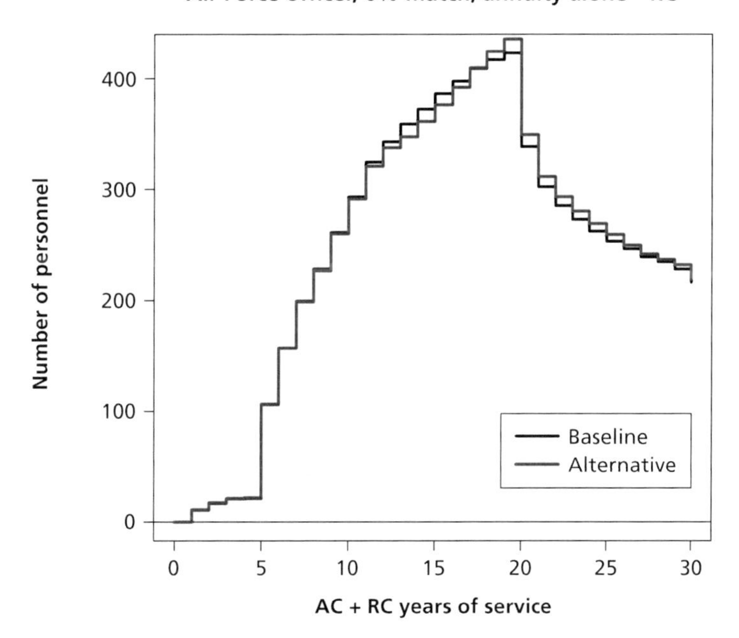

RAND *RR1022-A.9b*

Figure A.9c
Air Force AC Enlisted Retention and Prior-Active RC Enlisted Participation, 0-Percent DoD Match Rate, Partial Annuity/Partial Lump-Sum Choice

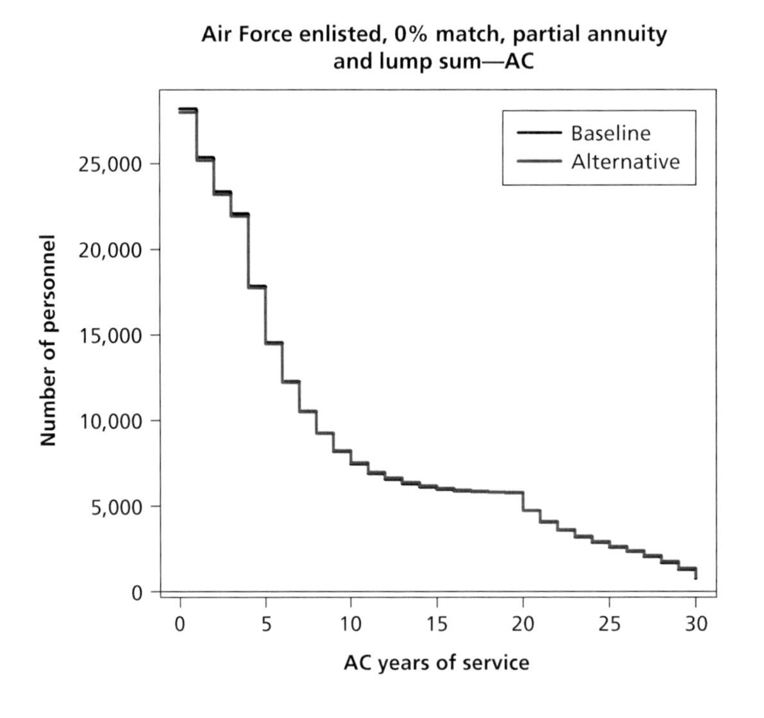

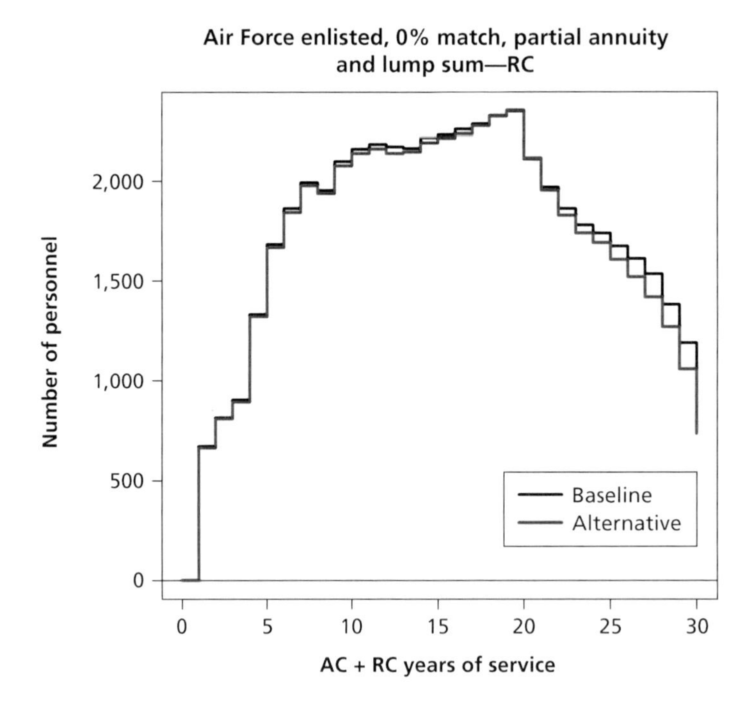

RAND *RR1022-A.9c*

Figure A.9d
Air Force AC Enlisted Retention and Prior-Active RC Enlisted Participation, 0-Percent DoD Match Rate, No Annuity/Full Lump-Sum Choice

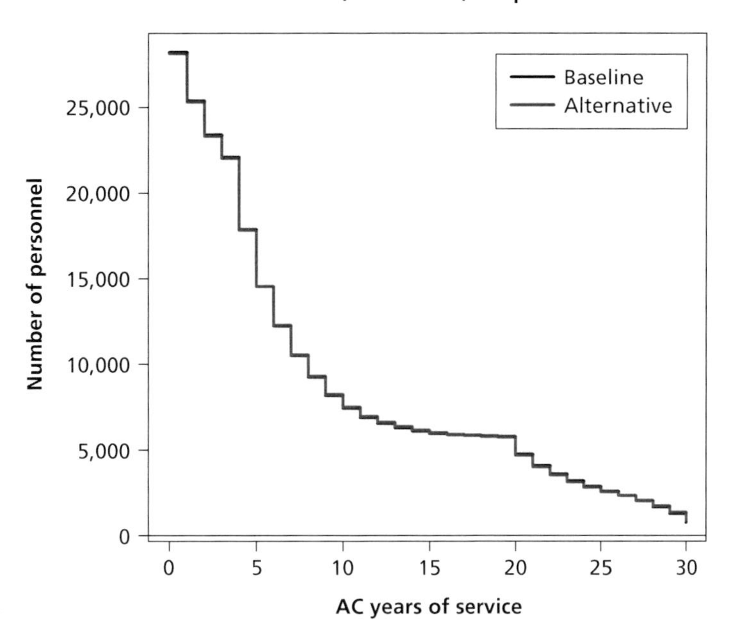

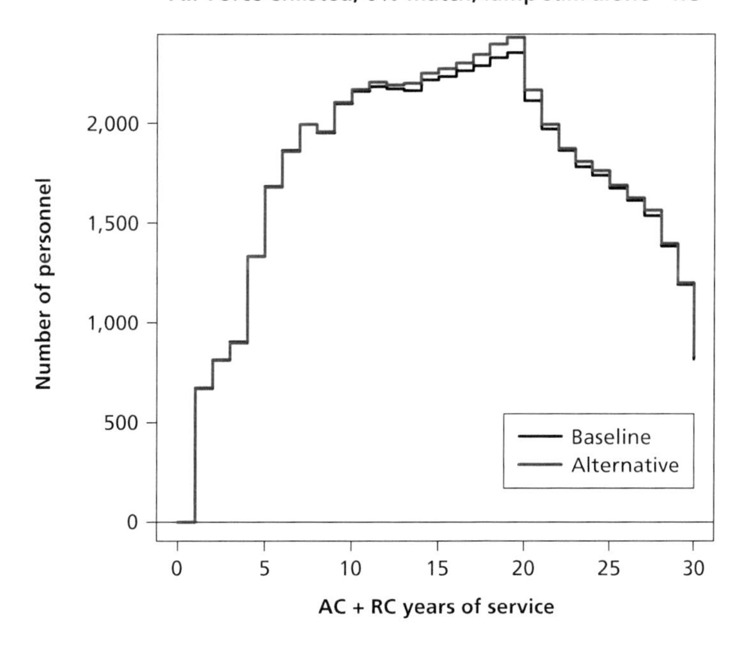

RAND *RR1022-A.9d*

Figure A.10a
Air Force AC Enlisted Retention and Prior-Active RC Enlisted Participation, 3-Percent DoD Match Rate, Annuity-Only Choice

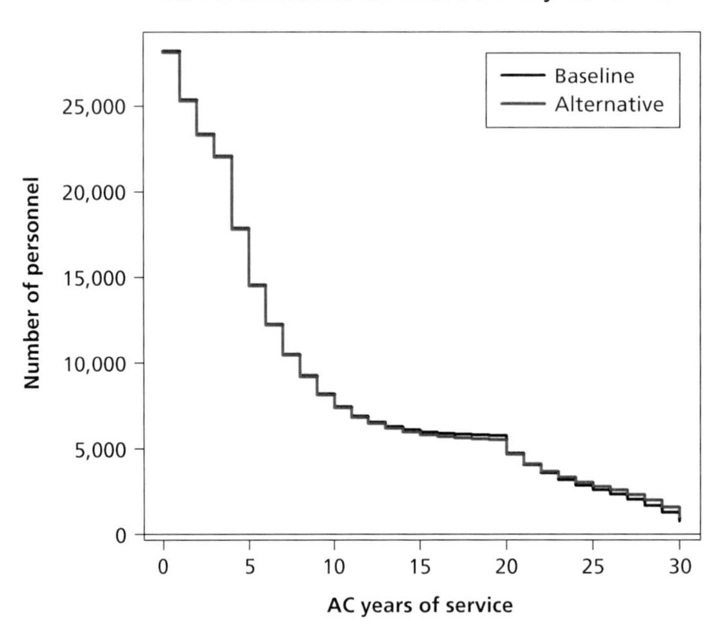

Air Force enlisted, 3% match, annuity alone—RC

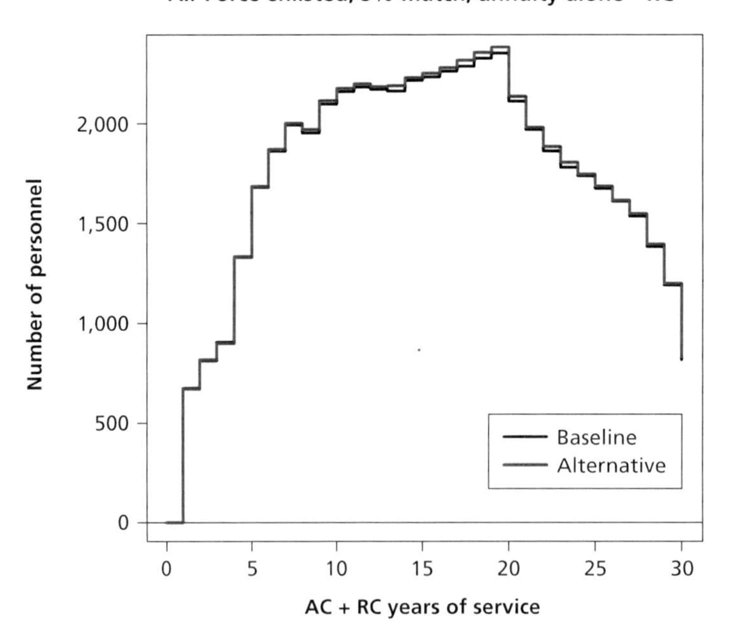

RAND *RR1022-A.10a*

Figure A.10b
Air Force AC Officer Retention and Prior-Active RC Officer Participation, 3-Percent DoD Match Rate, Annuity-Only Choice

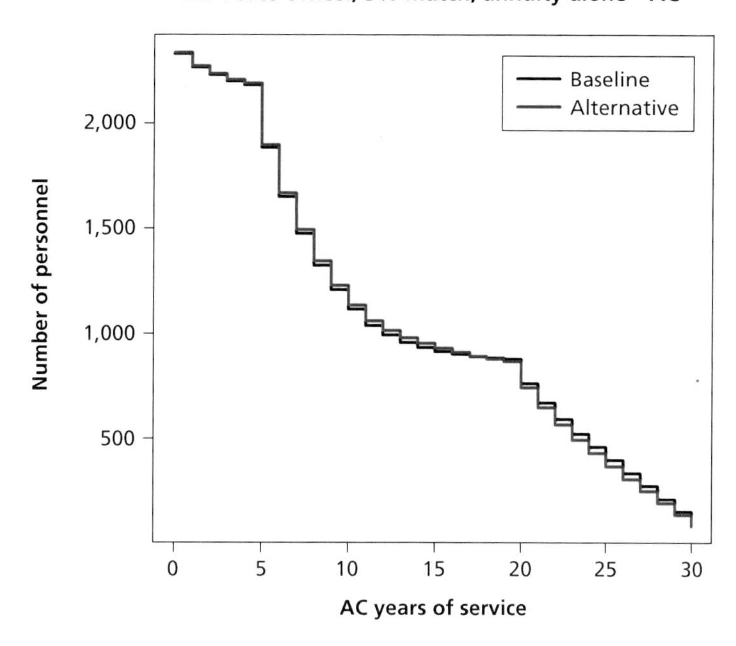

Air Force officer, 3% match, annuity alone—RC

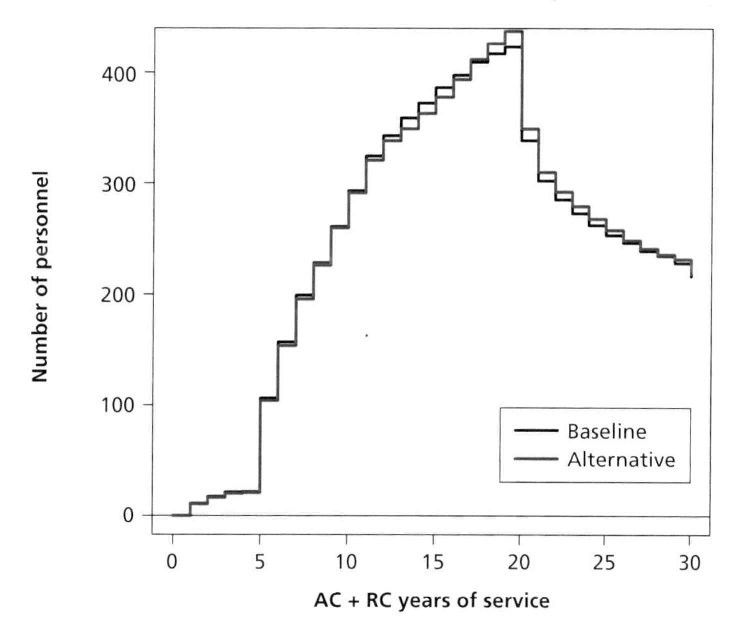

RAND *RR1022-A.10b*

Figure A.10c
Air Force AC Enlisted Retention and Prior-Active RC Enlisted Participation, 3-Percent DoD Match Rate, Partial Annuity/Partial Lump-Sum Choice

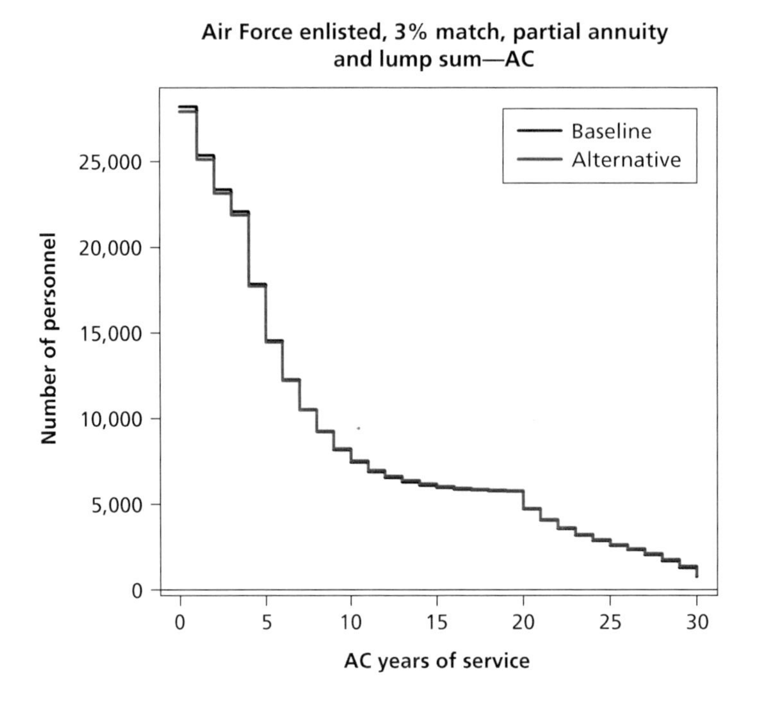

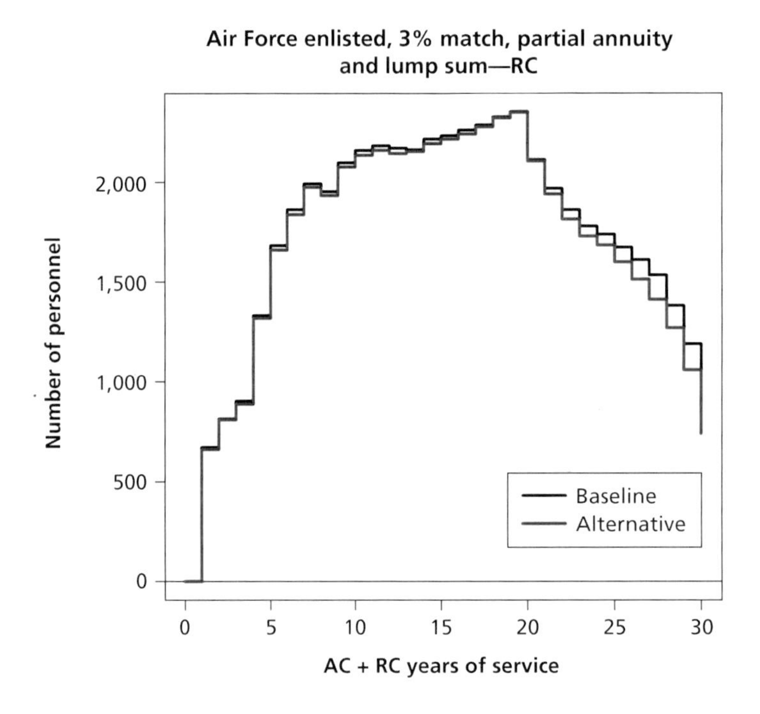

RAND *RR1022-A.10c*

Figure A.10d
Air Force AC Enlisted Retention and Prior-Active RC Enlisted Participation, 3-Percent DoD Match Rate, No Annuity/Full Lump-Sum Choice

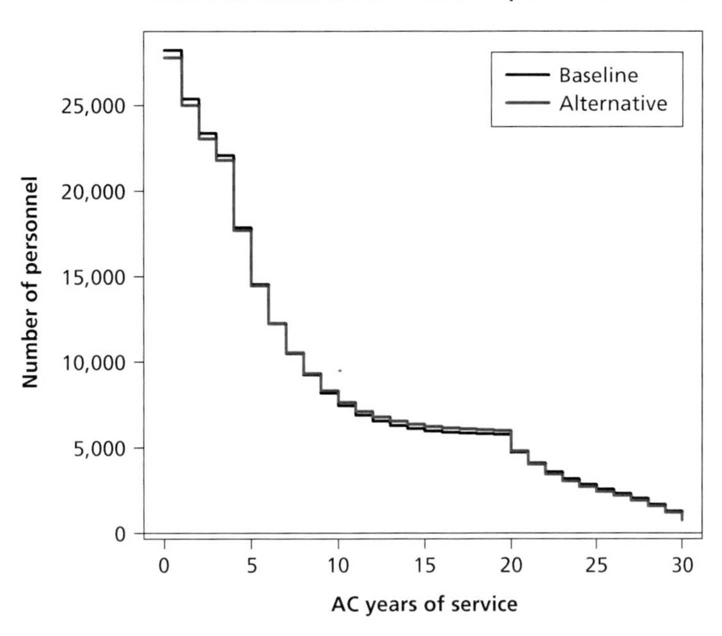

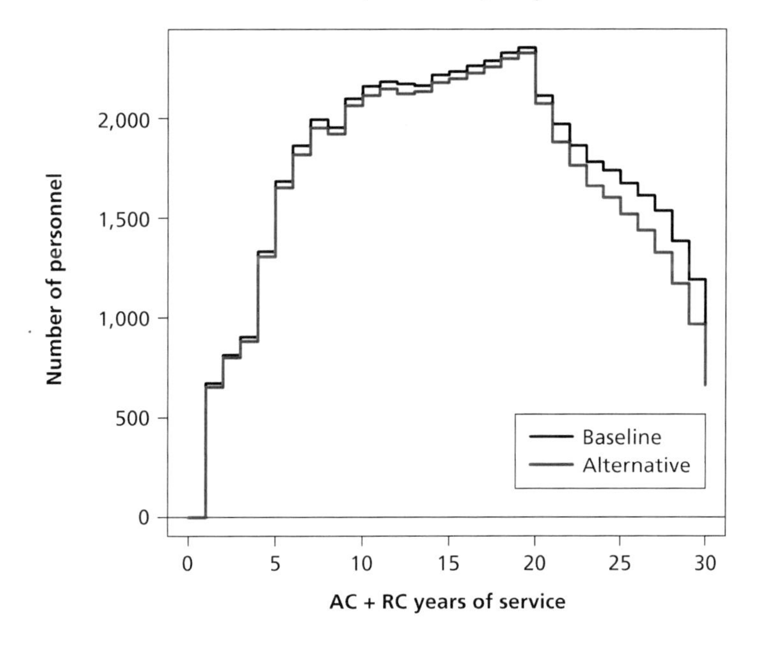

RAND *RR1022-A.10d*

Figure A.11a
Air Force AC Enlisted Retention and Prior-Active RC Enlisted Participation, 5-Percent DoD Match Rate, Annuity-Only Choice

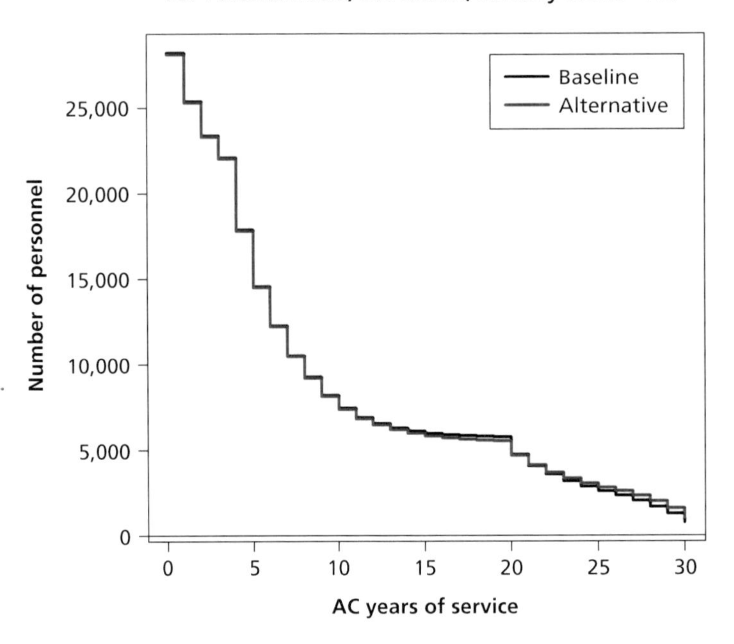

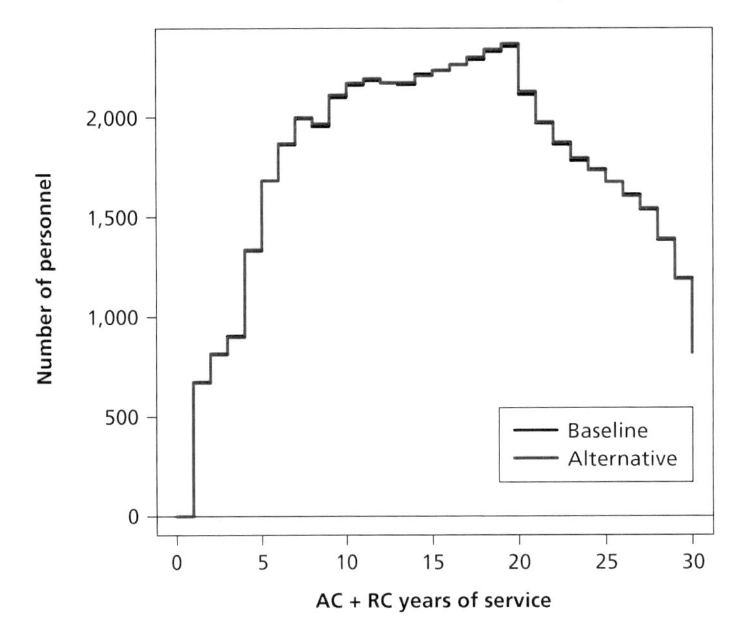

RAND *RR1022-A.11a*

Figure A.11b
Air Force AC Officer Retention and Prior-Active RC Officer Participation, 5-Percent DoD Match Rate, Annuity Only Choice

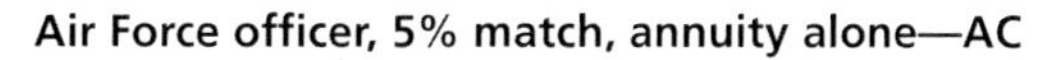

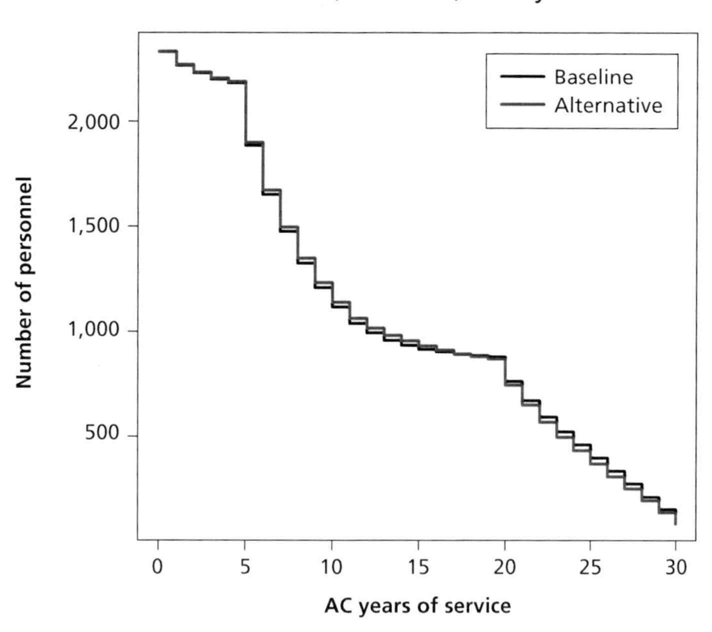

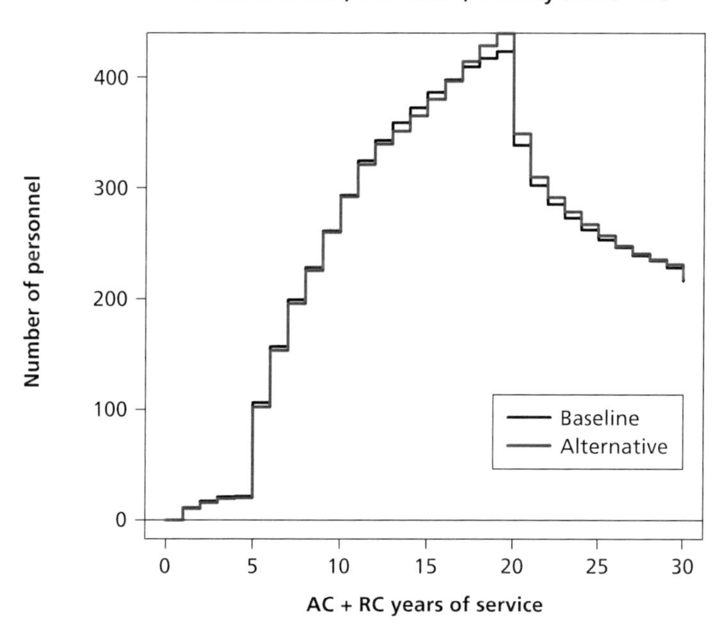

RAND *RR1022-A.11b*

Figure A.11c
Air Force AC Enlisted Retention and Prior-Active RC Enlisted Participation, 5-Percent DoD Match Rate, Partial Annuity/Partial Lump-Sum Choice

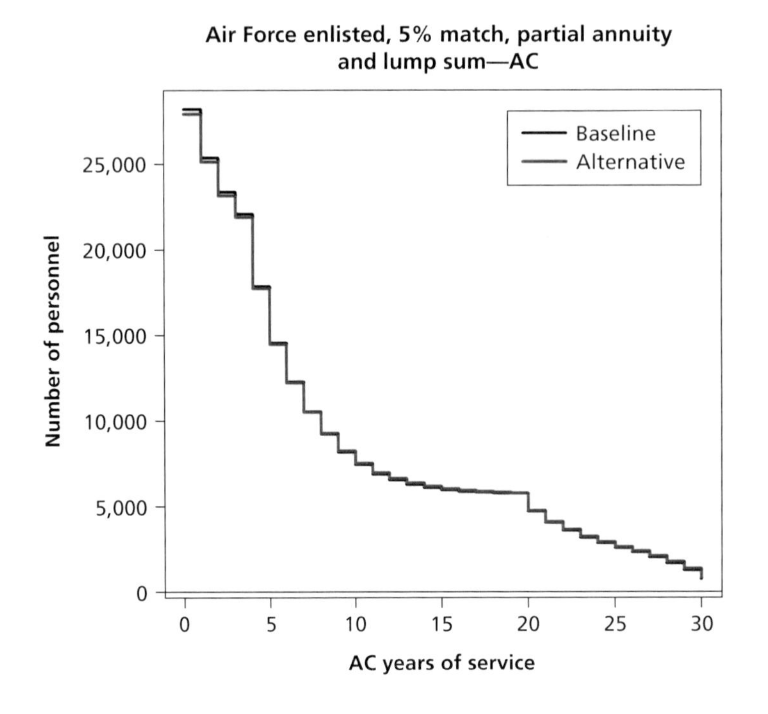

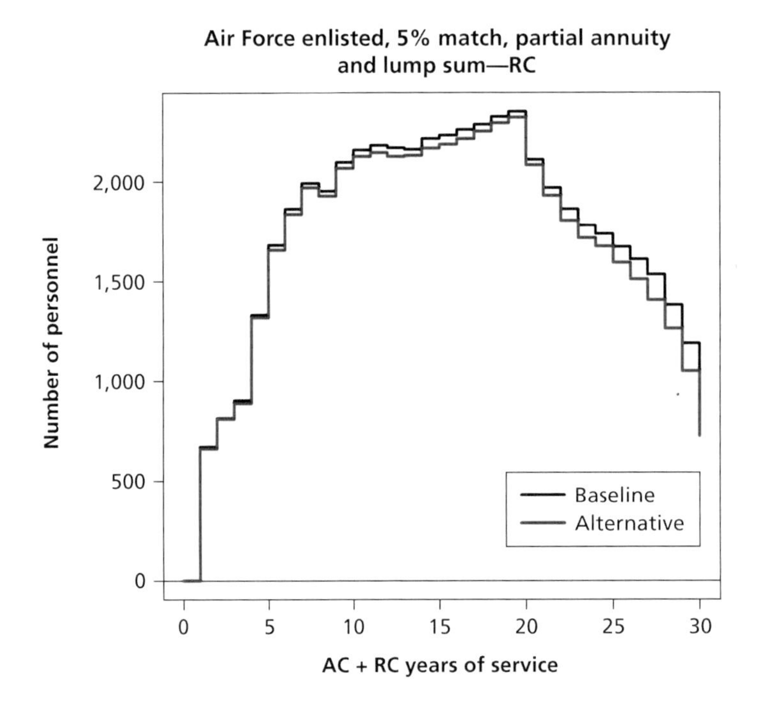

RAND *RR1022-A.11c*

Figure A.11d
Air Force AC Enlisted Retention and Prior-Active RC Enlisted Participation, 5-Percent DoD Match Rate, No Annuity/Full Lump-Sum Choice

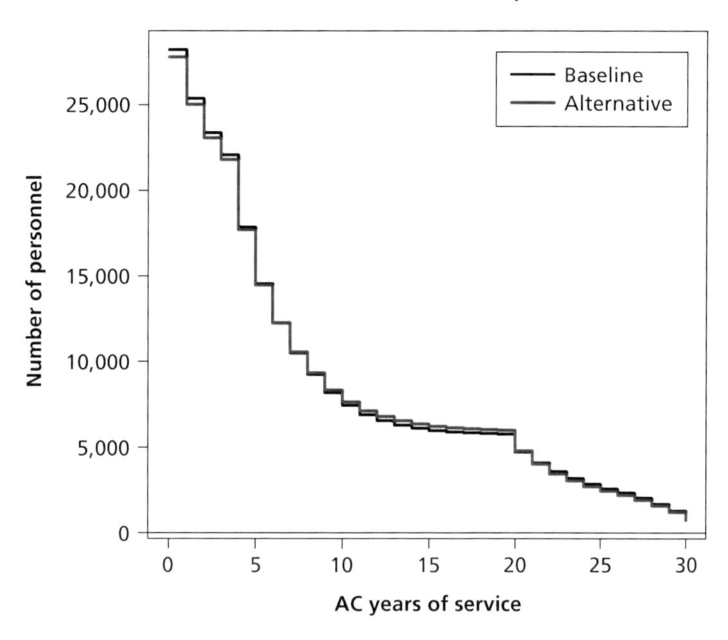

Air Force enlisted, 5% match, lump sum alone—RC

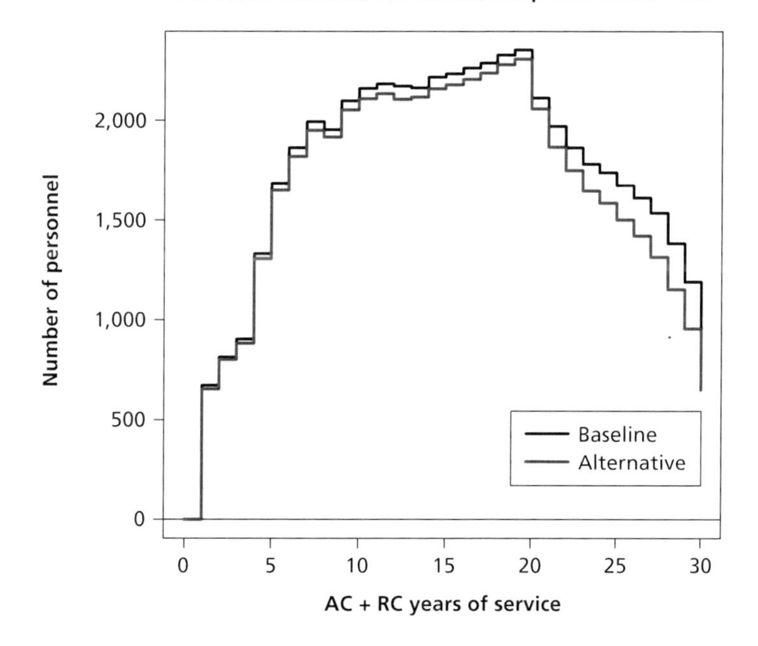

RAND *RR1022-A.11d*

APPENDIX B

Lump-Sum Formula

This appendix describes the lump-sum formulas for AC and RC members. AC members who are eligible to retire would have the option of (1) receiving the full annuity; (2) receiving a lump-sum payment along with a reduced annuity based on a multiplier of 1 percent up to and including age 66, and receiving the full annuity thereafter; or (3) receiving a larger lump-sum payment along with a full annuity that would begin at age 67—in this option there would be no annuity payments from the age of retirement up to and including age 66. Thus, AC members could choose to take all or part of their annuity between retirement age and age 67 as either a partial or full lump sum, and the AC lump-sum formula is given below. RC members would have a similar choice in the sense that, at the time of retirement from the RC, they could choose to receive the full annuity from ages 60 to 67; a partial annuity from ages 60 to 67 and a partial lump sum, based on the RC formula below, paid at the time of retirement from the RC; or a full lump sum, based on the RC formula below, paid at the time of retirement from the RC.

One way to specify the AC lump sum is to make it equivalent to the discounted present value of the annuity from the year of service at retirement until age 67. This is given by the formula:

$$\text{Equivalent lump sum} = M \times \text{YOS} \times \text{BP} = \left[\left(\frac{1-\beta^{67-\text{EntryAge}-YOS}}{1-\beta}\right)\right] \times \text{YOS} \times \text{BP} \qquad \text{B.1}$$

where:

YOS	=	years of service at retirement
BP	=	average of the highest three years of basic pay at retirement
M	=	defined benefit multiplier (2 percent for the full lump sum, or 1 percent for the partial lump sum)
β	=	discount factor equal to $1/(1 + r)$, where r is the personal discount rate.

Thus, the equivalent lump sum in Equation B.1 depends on a multiplier times YOS and the average of the highest three years of basic pay, where the multiplier depends on age, YOS, and the personal discount factor. We specify the exponent to β in the multiplier in terms of (67–Entry Age–YOS) rather than its equivalent (67–Retirement Age) so that we examine how the formula varies with changes in entry age, holding YOS fixed, and with changes in YOS, holding entry age fixed. Furthermore, we fix entry age in the DRM, so that retirement age is driven only by YOS in the DRM. This is important because we use the DRM to help establish the parameters of the formula we develop, as described below.

Figure B.1 shows a simulation of the equivalent lump sum to the annuity between retirement and age 67 by YOS, assuming the member entered at age 18 and had a discount factor of 0.90. We use the 2014 basic pay table and weight it with the 2007 grade-by-YOS distribution to compute average annual basic pay by YOS. We use the 2007 weights because retention increased after 2007 as a result of the Great Recession. The key point of Figure B.1 is that the equivalent AC lump sum increases with YOS, despite the fact that those who separate with more years of service also have fewer years until age 67. The increase in the equivalent lump sum with years of service occurs because its formula increases in YOS and in BP, which indirectly increases with YOS, and these increases offset the fact that the multiplier *M* (shown in parentheses in Equation B.1) declines with YOS because individuals with more years of service are older and closer to age 67. In addition, a key finding of Figure B.1 is that the equivalent lump sum is practically linear in YOS at retirement, despite the fact that the multiplier *M* is nonlinear in years at separation. In general, the relationship between the equivalent lump sum and YOS at retirement is nonlinear, but it is nearly linear over the range of YOS 20–30. We use this fact in developing a simplified version of the formula for the lump sum below.

Figure B.2 shows that those who enter at older ages, and therefore retire from the AC at older ages and have fewer years until "old age" retirement at age 67, have a lower equivalent AC lump sum, holding YOS constant. For example, a member who enters at age 24 (the purple line) and therefore separates at an older age has a lower lump sum at YOS 30 than one who enters at age 18 (the blue line) and separates at a younger age, though both lines slope

Figure B.1
Equivalent AC Lump Sum to Annuity Between Retirement Age and Age 67, by YOS at Retirement

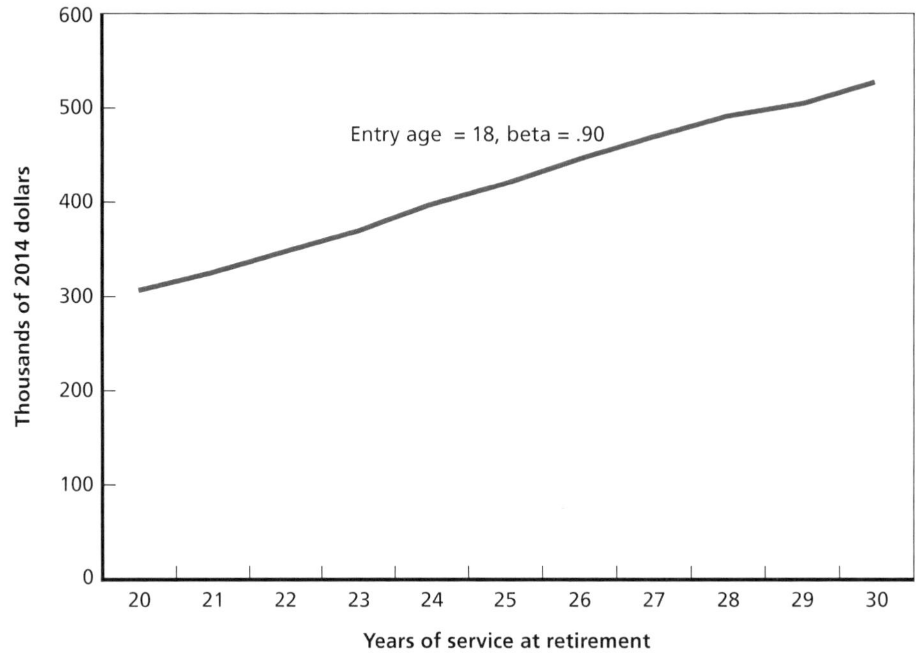

RAND *RR1022-B.1*

upward. This occurs because the multiplier, M, declines for those separating later (Figure B.3). The decline in the multiplier produces a small negative effect on the equivalent lump sum, which, overall, increases with YOS and BP, as in Figure B.2.

Two other observations are important regarding Figure B.3. First, it shows that the multiplier depends on the discount factor. Those with higher discount factors (lower discount rates) will have a higher multiplier and therefore a higher lump-sum equivalent. Second, the value of the multiplier is roughly between 0.20 and 0.30 for those entering at age 22 and having a higher discount factor, corresponding to officers, while the value is roughly between 0.15 and 0.20 for those entering at younger ages and having lower discount factors, corresponding to enlisted personnel.

The MCRMC pay and retirement working group asked us to develop a simplified formula for the lump sum that could be readily understood by service members, yet at the same time provide an amount that sustained retention incentives relative to the baseline, in conjunction with the other elements of the MCRMC reform package. If possible, the formula would be similar to the retirement benefit formula, which was familiar to service members. This meant that the formula would not depend on discount factor—and dropping the discount factor was feasible because, as seen, the equivalent lump sum was nearly linear in YOS at retirement. Still, there is some nonlinearity that becomes apparent when the age of retiring from the military is high and there are relatively few years remaining between separation and age 67. In this case, we adapt the linear approximation by making it a spline, as explained next.

Figure B.2
Equivalent Lump Sum to Annuity, by YOS at Retirement, Assuming Different Entry Ages

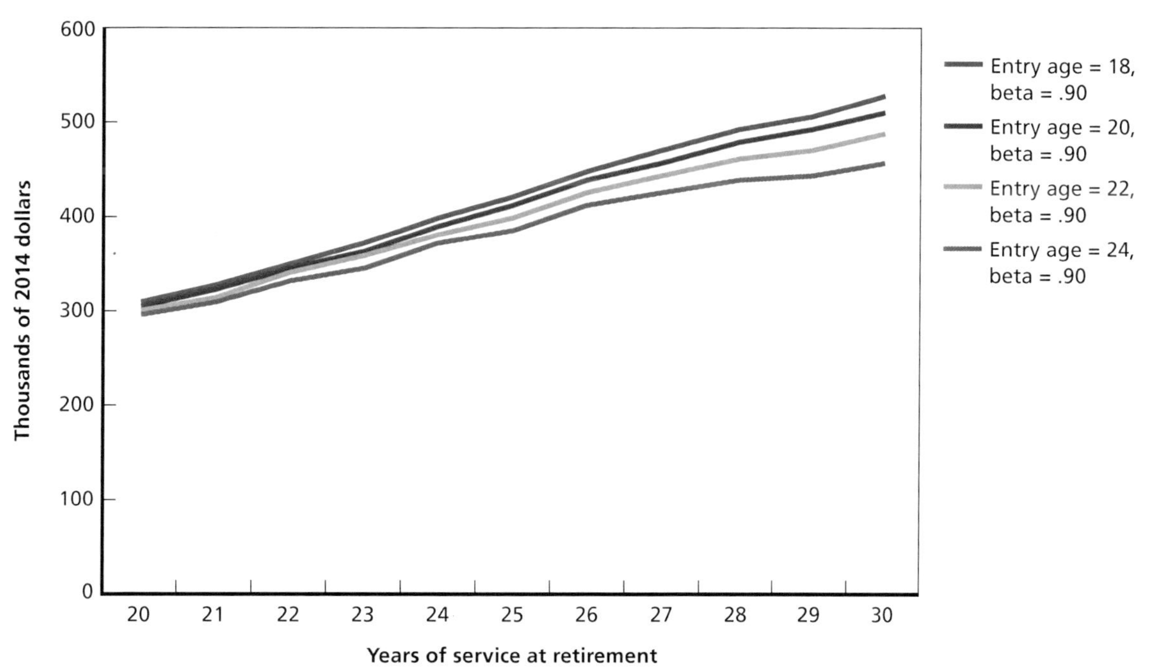

RAND *RR1022-B.2*

Figure B.3
Equivalent Lump-Sum Multiplier, by YOS at Retirement

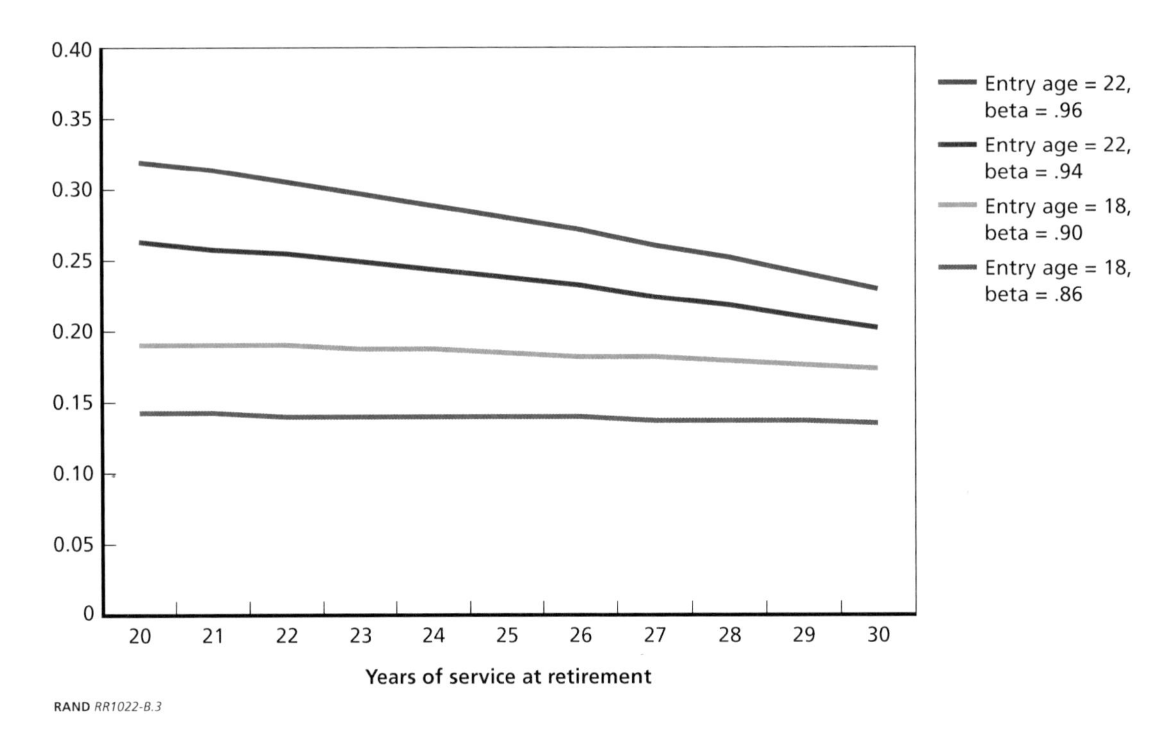

RAND *RR1022-B.3*

Equation B.2 shows the simplified formula we use to compute the AC lump sum:

$$\text{Simplified AC lump sum} = \hat{M} \times \text{YOS} \times \text{BP}, \tag{B.2}$$

where

$$\hat{M} = a - b\,\text{Age, if Age} < 55$$

$$\hat{M} = a - b\,\text{Age} - 2b\,(\text{Age} - 54)\text{, if age} \geq 55.$$

Here, Age is the age at retirement from the military. The values of a and b are the optimized fit values in the DRM, as discussed in Chapter Three and below. If age is less than 55, the multiplier decreases by b for each year of age. If age is 55 or greater, the multiplier decreases at a faster rate, but the intercept is effectively higher.[1]

For the RC, members can choose to receive an annuity between ages 60 and 67 or a partial or full lump sum, though if they choose a lump-sum option they receive it at retirement from the reserves (i.e., separation from the reserves with 20 or more creditable years of service), which can occur before or after age 60. We found that the following formula could be used for the lump sum:

$$\text{Simplified RC lump sum} = \tilde{M} \times \text{RCYOS} \times \text{BP}, \tag{B.3}$$

---

[1] The multiplier formula can be rewritten as $\hat{M} = (a + 2b\,(54)) - 3b\,\text{Age, if Age} \geq 55$.

where RCYOS is years of creditable service in the RC and where

$\tilde{M}$ = Maximum ($a - b$ (years to age 60), one month of the average highest three years of monthly basic pay), if Age < 60

$\tilde{M}$ = (Sum of annuity payments until Age 67), if Age ≥ 60

The formulas we developed were the result of a two-step process. The first step involved regression analyses to identify formulas that could replicate the natural lump-sum formula given in Equation B.1 in the case of the AC. We simulated data of the natural formulas by varying age of separation and assuming a fixed personal discount rate[2] and then estimated linear regression models for the multiplier for the AC and then the RC to identify the values of a and b.[3] Initially, in the case of the AC, the regression did not include a spline bend point at age 55. But we found that without the bend point, the estimated regression did a poor job of replicating the natural formula at ages 55 and above, reflecting the structure of the military pay table and the nonlinear multiplier in Equation B.1. We therefore altered the AC formula to include the spline, reestimated the linear regression model, and found that the estimated parameters with the spline formula did a much better job of replicating the natural formula. Similarly, we found that the linear RC formula worked well in most cases, but because there are many combinations of AC and RC service that qualify a member for RC retirement and many possible retirement ages, we found that in a handful of cases a simple formula could result in negative lump-sum values. To prevent this, we modified the formula so that an RC member at a minimum always received one month of the average monthly basic pay from the three highest years of basic pay. For those who retire from the RC when they are 60 or older, we found that the sum of the face value of the remaining annuity payments replicated the natural formula for this group better than the linear formula.

These estimated parameters from the regression analysis were input to the second step of the process. In policy simulations, we optimized the continuation pay multipliers for the AC and for the RC, the two parameters of the AC formula (a and b in Equation B.2), and the two parameters of the RC lump-sum formula (a and b in Equation B.3). We were required to optimize these parameters, rather than just use the parameters that emerged from our regressions, because the parameters and, therefore, the lump-sum amounts—together with the rest of the MCRMC reform package—must sustain retention relative to the baseline. That said, the estimated parameters from the regression analysis were the starting values for the estimation of the optimized parameters in the DRM, shown in Tables B.1 and B.2.

Thus, we have six parameters to optimize in the model. We optimize these parameters for each DC match case (0 percent, 3 percent, 5 percent) and for each lump-sum choice (all choose annuity, all choose partial annuity/partial lump sum, all choose full lump sum) using the two-step procedure described in Chapter Three. As explained in the text, to cut down on the number of simulations, we assume that the lump-sum choice in the RC is the same as in

[2] The rate selected was 10 percent, at the request of the sponsor. This was a real rate (adjusted for inflation).

[3] The estimated linear regression for the AC was: Multiplier_AC = 0.3470 – (0.0034 × AgeAtSeparation) – (0.0064 × AgeAtSep55), where AgeAtSep55 is the maximum of zero and AgeAtSeparation – 54, with standard errors of 0.004 for the intercept, 0.00008 for AgeAtSeparation, and 0.00017 for AgeAtSept55. The estimated linear regression for the RC was: Multiplier_RC = 0.0963 – (0.005 × YearsUntilAge60), with standard errors of 0.00017 for the intercept and 0.00002 for YearsUntilAge60.

Table B.1
Optimized Full Lump-Sum Parameters for AC and RC, by Service

	DoD DC Match: 0%				DoD DC Match: 3%				DoD DC Match: 5%			
	AC		RC		AC		RC		AC		RC	
	a	*b*	*a*	*b*	*a*	*b*	*a*	*b*	*a*	*b*	*a*	*b*
Enlisted												
Army	0.322	0.0037	0.086	0.0058	0.277	0.0027	0.092	0.0065	0.278	0.0027	0.102	0.0077
Navy	0.319	0.0031	0.130	0.0202	0.283	0.0022	0.097	0.0050	0.254	0.0017	0.157	0.1960
Air Force	0.297	0.0030	0.104	0.0069	0.287	0.0028	0.163	0.0429	0.293	0.0029	0.108	0.0070
Marine Corps	0.322	0.0031	0.093	0.0035	0.322	0.0031	0.085	0.0004	0.316	0.0030	0.134	0.0166
Officer												
Army	0.472	0.0051	0.168	0.0269	0.449	0.0046	0.121	0.0095	0.453	0.0046	0.127	0.0094
Navy	0.447	0.0042	0.137	0.0478	0.439	0.0041	0.135	0.6073	0.505	0.0053	0.137	0.0169
Air Force	0.472	0.0049	0.168	0.0261	0.429	0.0040	0.145	0.0168	0.477	0.0048	0.157	0.0161
Marine Corps	0.470	0.0051	0.141	0.0160	0.434	0.0044	0.127	0.0167	0.447	0.0045	0.136	0.0150

Table B.2
Optimized Partial Annuity/Partial Lump-Sum Parameters for AC and RC, by Service

	DoD DC Match: 0%				DoD DC Match: 3%				DoD DC Match: 5%			
	AC		RC		AC		RC		AC		RC	
	a	*b*	*a*	*b*	*a*	*b*	*a*	*b*	*a*	*b*	*a*	*b*
Enlisted												
Army	0.171	0.0020	0.055	0.0037	0.148	0.0015	0.050	0.004	0.163	0.0019	0.057	0.0044
Navy	0.140	0.0011	0.072	0.0150	0.170	0.0017	0.051	0.003	0.171	0.0018	0.089	0.0306
Air Force	0.178	0.0020	0.044	0.0029	0.178	0.0020	0.048	0.003	0.180	0.0019	0.051	0.0033
Marine Corps	0.181	0.0019	0.048	0.0000	0.180	0.0019	0.044	0.000	0.231	0.0023	0.050	0.0024
Officer												
Army	0.254	0.0030	0.111	0.0273	0.226	0.0024	0.101	0.0264	0.231	0.0023	0.099	0.0211
Navy	0.228	0.0023	0.053	0.0084	0.185	0.0013	0.074	0.0148	0.234	0.0022	0.053	0.0035
Air Force	0.218	0.0021	0.127	0.0342	0.181	0.0013	0.028	0.0017	0.200	0.0016	0.030	0.0028
Marine Corps	0.207	0.0020	0.066	0.0093	0.204	0.0019	0.083	0.0527	0.219	0.0021	0.086	0.0262

the AC. For example, if we assume full lump sum in the AC, we make the same assumption for the RC.

The optimized lump-sum parameters are shown in Table B.1 for the full lump sum and Table B.2 for the partial annuity/partial lump sum. For enlisted personnel, the optimized value of *a* for the full lump sum in Table B.1 is about the same for each service, regardless of the DC matching rate, and ranges from 25 to 33 percent of the average of the highest three years of basic pay times YOS. The optimized values of *a* for the full lump sum are larger for officers, ranging from about 43 percent to about 51 percent. The higher optimized values for officers reflects the fact that they have lower personal discount rates, as estimated by our model, so they discount the annuity at a lower rate, which implies that it has a higher present value. Consequently, the lump sum must be larger to produce the same retention effect as the annuity.

The parameter *b* times the average of the highest three years of basic pay times YOS is the amount that the lump sum is reduced for those retiring at older ages, and therefore with fewer years until reaching age 67. For enlisted personnel, the optimized *b* is generally around 0.3 percent, while for officers it is generally around 0.5 percent. Thus, an enlisted member who is a year older and a year closer to age 67 has a 0.3 percentage point reduction, while an officer has a 0.5 percentage point reduction, holding the average of the highest three years of basic pay and total YOS equal.

In the case of the RC, the lump-sum payment equals a multiple of the average of the highest three years of basic pay times YOS, where the multiplier again depends on parameters *a* and *b*. For enlisted, the optimized RC full lump-sum parameter *a* ranges from 8 percent to 16 percent of the highest three years of annual RC pay times YOS. For officers, parameter *a* ranges from 13 percent to 17 percent. The parameter *b* ranges from 0.2 percent to 19.6 percent for enlisted and from 0.4 percent to 60.7 percent for officers.

On the basis of these optimized parameters for the lump sums, the MCRMC selected the values in Table 2.3. The selected values are the same across service and for officers and enlisted personnel. The selected values are generally equal or greater than the optimized values for enlisted personnel but are smaller than the optimized values for officers. This implies that few officers, given the personal discount rates for officers estimated in the DRM, will select the lump-sum option, and most will prefer the annuity choice. In contrast, given the estimated personal discount rates for enlisted personnel, they will tend to be indifferent or prefer the lump-sum options.

Table B.2 shows the partial annuity/partial lump-sum multipliers that sustain retention relative to the baseline when offered in conjunction with the DB and DC plans and the continuation pays. Since the partial annuity is one-half of the full annuity during the period between retirement and age 67, it would seem logical that the partial lump sum would be one-half of the full lump sum, so the multiplier parameters should also be about one-half. Comparison of Table B.2 with Table B.1 shows that the partial multipliers are about one-half or a bit more than one-half in some cases. The reason that they are not exactly one-half is that the values in the tables are optimized to sustain retention. Some small deviations between the baseline retention and retention under the reform alternative can change the optimized values. Still, as expected, they are roughly one-half of those reported in Table B.1. The MCRMC selected parameters for the partial lump sum in Table 2.3 that are one-half of the ones for the full lump sum.

Applying the formula for the simplified lump sum in Equation B.2 gives the value of the multiplier, which depends on years of service at retirement. Figure B.4 shows the value of the

Figure B.4
Enlisted Simplified Lump-Sum Multiplier with Optimized Parameters

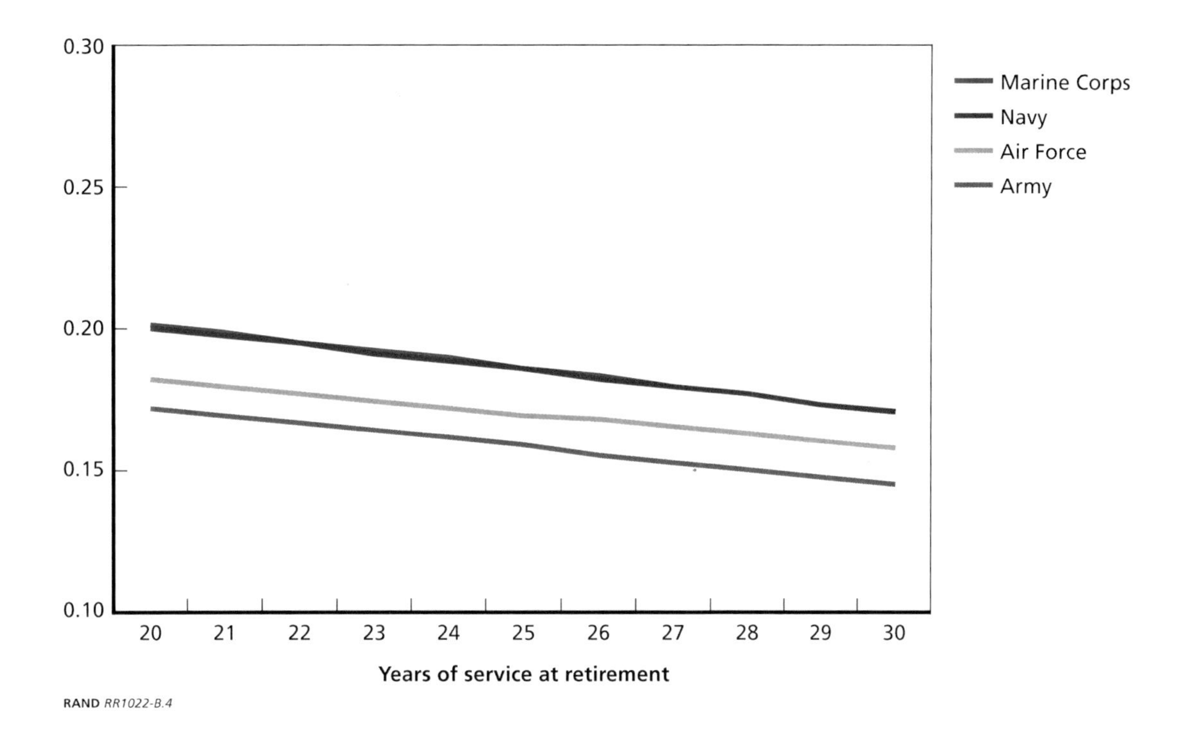

RAND *RR1022-B.4*

simplified AC multiplier for enlisted personnel, using the optimized values of *a* and *b*, by YOS at retirement. The value of the multiplier using optimized parameters for the full lump sum is between 0.15 and 0.20, just as in the case of the equivalent lump sum in Figure B.3. Figure B.5 shows the value of the multiplier for officers, using optimized values of *a* and *b*. The value lies between 0.2 and 0.3. The fact that the multipliers using the optimized values are close to the multipliers we get from the formula for the lump-sum equivalent as well as the regression results used in the first step gives us confidence that the simplified formula is a good approximation to the "true" formula in Equation B.1. Adding to our confidence is the fact that we find excellent fits of the retention profiles relative to the baseline. That is, the optimized parameters do a good job of "buying back" retention.

Figure B.5
Officer Simplified Lump-Sum Multiplier with Optimized Parameters

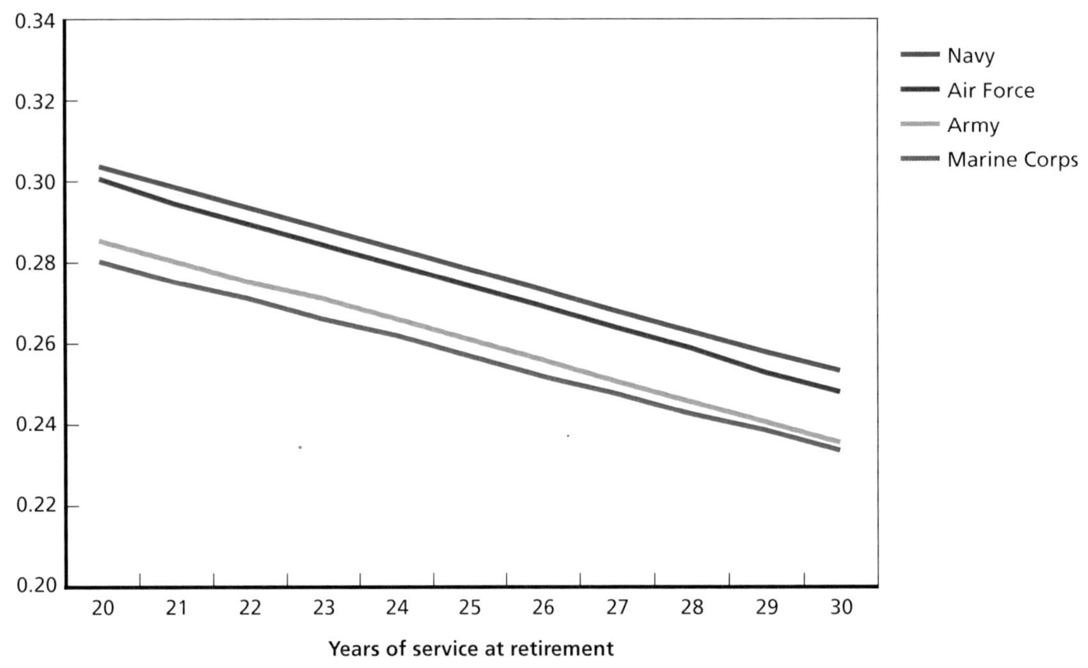

RAND *RR1022-B.5*

References

Asch, Beth J., James Hosek, and Michael G. Mattock, *A Policy Analysis of Reserve Retirement Reform*, Santa Monica, Calif.: RAND Corporation, MG-378-OSD, 2013. As of December 10, 2013:
http://www.rand.org/pubs/monographs/MG378.html

Asch, Beth J., James Hosek, and Michael G. Mattock, *Toward Meaningful Compensation Reform: Research in Support of DoD's Review of Military Compensation*, Santa Monica, Calif.: RAND Corporation, RR-501-OSD, 2014. As of March 23, 2015:
http://www.rand.org/pubs/research_reports/RR501.html

Asch, Beth J., Michael G. Mattock, and James Hosek, *A New Tool for Assessing Workforce Management Policies Over Time: Extending the Dynamic Retention Model*, Santa Monica, Calif.: RAND Corporation, RR-113-OSD, 2013. As of December 10, 2013:
http://www.rand.org/pubs/research_reports/RR113.html

Defense Advisory Committee on Military Compensation, *Completing the Transition to an All-Volunteer Force: Report of the Defense Advisory Committee on Military Compensation,* Arlington, Va., April 2006.

Defense Science Board, *The Defense Science Board Task Force on Human Resources Strategy*, Washington, D.C.: Office of the Under Secretary of Defense for Acquisition, Technology, and Logistics, February 2000.

Goldberg, Matthew S., "A Survey of Enlisted Retention: Models and Findings," *The Ninth Quadrennial Review of Military Compensation, Volume III, Chapter II*, Washington, D.C., 2002. As of March 2, 2015:
http://prhome.defense.gov/RFM/MPP/Reports

Gotz, Glenn A., "Comment on 'The Dynamics of Job Separation: The Case of Federal Employees,'" *Journal of Applied Econometrics*, Vol. 5, No. 3, July–September 1990, pp. 263–268.

Henning, Charles A., *Military Retirement Reform: A Review of Proposals and Options for Congress*, Washington D.C.: Congressional Research Service, 2011.

Madrian, Brigitte, and Dennis Shea, "The Power of Suggestion: Inertia in 401(k) Participation and Savings Behavior," *Quarterly Journal of Economics*, Vol. 66, No. 4, pp. 1149–1188, 2001.

Mattock, Michael G., James Hosek, and Beth J. Asch, *Reserve Participation and Cost Under a New Approach to Reserve Compensation*, Santa Monica, Calif.: RAND Corporation, MG-1153-OSD, 2012. As of December 10, 2013:
http://www.rand.org/pubs/monographs/MG1153.html

MCRMC—*See* Military Compensation and Retirement Modernization Commission.

Military Compensation and Retirement Modernization Commission, *Report of the Military Compensation and Retirement Modernization Commission, Final Report*, Washington D.C., January 2015. As of February 27, 2015:
http://www.mcrmc.gov/public/docs/report/MCRMC-FinalReport-29JAN15-LO.pdf

U.S. Department of Defense, *Report of the Tenth Quadrennial Review of Military Compensation, Volume I: Cash Compensation*, Washington, D.C., February 2008a.

U.S. Department of Defense, *Report of the Tenth Quadrennial Review of Military Compensation, Volume II: Deferred and Noncash Compensation*, Washington, D.C., July 2008b.

U.S. Department of Defense, *Concepts for Modernizing Military Retirement*, Washington, D.C., March 2014.